Springer Theses

Recognizing Outstanding Ph.D. Research

Aims and Scope

The series "Springer Theses" brings together a selection of the very best Ph.D. theses from around the world and across the physical sciences. Nominated and endorsed by two recognized specialists, each published volume has been selected for its scientific excellence and the high impact of its contents for the pertinent field of research. For greater accessibility to non-specialists, the published versions include an extended introduction, as well as a foreword by the student's supervisor explaining the special relevance of the work for the field. As a whole, the series will provide a valuable resource both for newcomers to the research fields described, and for other scientists seeking detailed background information on special questions. Finally, it provides an accredited documentation of the valuable contributions made by today's younger generation of scientists.

Theses are accepted into the series by invited nomination only and must fulfill all of the following criteria

- They must be written in good English.
- The topic should fall within the confines of Chemistry, Physics, Earth Sciences, Engineering and related interdisciplinary fields such as Materials, Nanoscience, Chemical Engineering, Complex Systems and Biophysics.
- The work reported in the thesis must represent a significant scientific advance.
- If the thesis includes previously published material, permission to reproduce this must be gained from the respective copyright holder.
- They must have been examined and passed during the 12 months prior to nomination.
- Each thesis should include a foreword by the supervisor outlining the significance of its content.
- The theses should have a clearly defined structure including an introduction accessible to scientists not expert in that particular field.

More information about this series at http://www.springer.com/series/8790

Tim Baarslag

Exploring the Strategy Space of Negotiating Agents

A Framework for Bidding, Learning
and Accepting in Automated Negotiation

Nominated as an outstanding PhD thesis by
Delft University of Technology, The Netherlands

 Springer

Author
Dr. Tim Baarslag
Interactive Intelligence Group, Department
    of Intelligent Systems, Faculty
    of Electrical Engineering, Mathematics
    and Computer Science
Delft University of Technology
Delft
The Netherlands

Supervisors
Prof. Catholijn M. Jonker
Interactive Intelligence Group, Department
    of Intelligent Systems, Faculty
    of Electrical Engineering, Mathematics
    and Computer Science
Delft University of Technology
Delft
The Netherlands

Dr. Koen V. Hindriks
Interactive Intelligence Group, Department
    of Intelligent Systems, Faculty
    of Electrical Engineering, Mathematics
    and Computer Science
Delft University of Technology
Delft
The Netherlands

Springer Theses
ISBN 978-3-319-80305-0   ISBN 978-3-319-28243-5   (eBook)
DOI 10.1007/978-3-319-28243-5

Printed on acid-free paper

This Springer imprint is published by SpringerNature
The registered company is Springer International Publishing AG Switzerland

Parts of this thesis have been published in the following articles:

The chapters in this thesis are based on publications in scientific journals and/or peer-reviewed conference proceedings, including the appendices. At the beginning of every chapter, we specify where parts have already been published. The full list is given below.

Journal Papers

Tim Baarslag, Mark J.C. Hendrikx, Koen V. Hindriks, and Catholijn M. Jonker. Learning about the opponent in automated bilateral negotiation: a comprehensive survey of opponent modeling techniques. *Autonomous Agents and Multi-Agent Systems*, pages 1–50, 2015

Raz Lin, Sarit Kraus, Tim Baarslag, Dmytro Tykhonov, Koen V. Hindriks, and Catholijn M. Jonker. Genius: An integrated environment for supporting the design of generic automated negotiators. *Computational Intelligence*, 30(1):48–70, 2014

Tim Baarslag, Katsuhide Fujita, Enrico H. Gerding, Koen V. Hindriks, Takayuki Ito, Nicholas R. Jennings, Catholijn M. Jonker, Sarit Kraus, Raz Lin, Valentin Robu, and Colin R. Williams. Evaluating practical negotiating agents: Results and analysis of the 2011 international competition. *Artificial Intelligence*, 198:73–103, May 2013

Tim Baarslag, Koen V. Hindriks, and Catholijn M. Jonker. Effective acceptance conditions in real-time automated negotiation. *Decision Support Systems*, 60:68–77, Apr 2014

Conference Papers

Tim Baarslag, Alexander S.Y. Dirkzwager, Koen V. Hindriks, and Catholijn M. Jonker. The significance of bidding, accepting and opponent modeling in automated negotiation. In *21st European Conference on Artificial Intelligence, volume 263 of Frontiers in Artificial Intelligence and Applications*, pages 27–32, 2014

Tim Baarslag, Mark J.C. Hendrikx, Koen V. Hindriks, and Catholijn M. Jonker. Predicting the performance of opponent models in automated negotiation. In *International Joint Conferences on Web Intelligence (WI) and Intelligent Agent Technologies (IAT), 2013 IEEE/WIC/ACM*, volume 2, pages 59–66, Nov 2013

Tim Baarslag and Koen V. Hindriks. Accepting optimally in automated negotiation with incomplete information. In *Proceedings of the 2013 International Conference on Autonomous Agents and Multi-agent Systems*, AAMAS '13, pages 715–722, Richland, SC, 2013. International Foundation for Autonomous Agents and Multiagent Systems

Tim Baarslag. Designing an automated negotiator: Learning what to bid and when to stop. In *Proceedings of the 2013 International Conference on Autonomous Agents and Multi-agent Systems*, AAMAS '13, pages 1419–1420, Richland, SC, 2013. International Foundation for Autonomous Agents and Multiagent Systems

Book Chapters

Tim Baarslag. *What to Bid and When to Stop*. Dissertation, Delft University of Technology, Sep 2014

Tim Baarslag, Koen V. Hindriks, Mark J.C. Hendrikx, Alex S.Y. Dirkzwager, and Catholijn M. Jonker. Decoupling negotiating agents to explore the space of negotiation strategies. In Ivan Marsa-Maestre, Miguel A. Lopez-Carmona, Takayuki Ito, Minjie Zhang, Quan Bai, and Katsuhide Fujita, editors, *Novel Insights in Agent-based Complex Automated Negotiation, volume 535 of Studies in Computational Intelligence*, pages 61–83. Springer, Japan, 2014

Tim Baarslag, Koen V. Hindriks, and Catholijn M. Jonker. Acceptance conditions in automated negotiation. In Takayuki Ito, Minjie Zhang, Valentin Robu, and Tokuro Matsuo, editors, *Complex Automated Negotiations: Theories, Models, and Software Competitions*, volume 435 of *Studies in Computational Intelligence*, pages 95–111. Springer Berlin Heidelberg, 2013

Tim Baarslag, Koen V. Hindriks, and Catholijn M. Jonker. A tit for tat negotiation strategy for real-time bilateral negotiations. In Takayuki Ito, Minjie Zhang, Valentin Robu, and Tokuro Matsuo, editors, *Complex Automated Negotiations: Theories, Models, and Software Competitions*, volume 435 of *Studies in Computational Intelligence*, pages 229–233. Springer Berlin Heidelberg, 2013

Katsuhide Fujita, Takayuki Ito, Tim Baarslag, Koen V. Hindriks, Catholijn M. Jonker, Sarit Kraus, and Raz Lin. The second automated negotiating agents competition (ANAC 2011). In Takayuki Ito, Minjie Zhang, Valentin Robu, and Tokuro Matsuo, editors, *Complex Automated Negotiations: Theories, Models, and Software Competitions*, volume 435 of *Studies in Computational Intelligence*, pages 183–197. Springer Berlin Heidelberg, 2013

Reyhan Aydoğan, Tim Baarslag, Koen V. Hindriks, Catholijn M. Jonker, and Pınar Yolum. Heuristic-based approaches for CP-nets in negotiation. In Takayuki Ito, Minjie Zhang, Valentin Robu, and Tokuro Matsuo, editors, *Complex Automated Negotiations: Theories, Models, and Software Competitions*, volume 435 of *Studies in Computational Intelligence*, pages 113–123. Springer Berlin Heidelberg, 2013

Tim Baarslag, Koen V. Hindriks, Catholijn M. Jonker, Sarit Kraus, and Raz Lin. The first automated negotiating agents competition (ANAC 2010). In Takayuki Ito, Minjie Zhang, Valentin Robu, Shaheen Fatima, and Tokuro Matsuo, editors, *New Trends in Agent-based Complex Automated Negotiations*, volume 383 of *Studies in Computational Intelligence*, pages 113–135, Berlin, Heidelberg, 2012. Springer-Verlag

Tim Baarslag, Mark J.C. Hendrikx, Koen V. Hindriks, and Catholijn M. Jonker. Measuring the performance of online opponent models in automated bilateral negotiation. In Michael Thielscher and Dongmo Zhang, editors, *AI 2012: Advances in Artificial Intelligence*, volume 7691 of *Lecture Notes in Computer Science*, pages 1–14. Springer Berlin Heidelberg, 2012

Tim Baarslag, Koen V. Hindriks, and Catholijn M. Jonker. Towards a quantitative concession-based classification method of negotiation strategies. In David Kinny, Jane Yung-jen Hsu, Guido Governatori, and Aditya K. Ghose, editors, *Agents in Principle, Agents in Practice*, volume 7047 of *Lecture Notes in Computer Science*, pages 143–158, Berlin, Heidelberg, 2011. Springer Berlin Heidelberg

Workshop Papers

Tim Baarslag, Rafik Hadfi, Koen V. Hindriks, Takayuki Ito, and Catholijn M. Jonker. Optimal non-adaptive concession strategies with incomplete information. In *Proceedings of The Seventh International Workshop on Agent-based Complex Automated Negotiations (ACAN 2014)*, 2014

Tim Baarslag. Accepting optimally in automated negotiation with incomplete information. In *Proceedings of the 25th Benelux Conference on Artificial Intelligence*, 2013

Tim Baarslag, Koen V. Hindriks, Mark J.C. Hendrikx, Alex S.Y. Dirkzwager, and Catholijn M. Jonker. Decoupling negotiating agents to explore the space of negotiation strategies. In *Proceedings of The Fifth International Workshop on Agent-based Complex Automated Negotiations (ACAN 2012)*, 2012

Tim Baarslag, Koen V. Hindriks, Mark J.C. Hendrikx, Alex S.Y. Dirkzwager, and Catholijn M. Jonker. Decoupling negotiating agents to explore the space of negotiation strategies. In *Proceedings of the 24th Benelux Conference on Artificial Intelligence*, 2012

Tim Baarslag, Koen V. Hindriks, and Catholijn M. Jonker. Acceptance conditions in automated negotiation. In Patrick De Causmaecker, Joris Maervoet, Tommy Messelis, Katja Verbeeck, and Tim Vermeulen, editors, *Proceedings of the 23rd Benelux Conference on Artificial Intelligence*, pages 363–365, 2011

Tim Baarslag, Koen V. Hindriks, and Catholijn M. Jonker. Acceptance conditions in automated negotiation. In *Proceedings of ICT.Open 2011*, 2011

For my parents

Supervisor's Foreword

I first met Tim in 2010, when he started as a Ph.D. student in the Interactive Intelligence Group of the Faculty of Electrical Engineering, Mathematics and Computer Science at Delft University of Technology. As head of the group, and as Tim's promoter, we interacted in weekly face-to-face meetings during the writing of his thesis and at various conferences and events. In these four years, I came to know Tim as a hardworking individual who is full of initiative. Tim has shown a great talent for formulating a research agenda and for recognizing the potential use of mathematical tools. His mathematical insights enabled him to independently come up with a number of tested theories on optimal strategies for negotiation, which outperform all existing mechanisms. As a result, his Ph.D. dissertation has yielded important practical and theoretical insights into the design of automated negotiators.

The dissertation was assessed by an international committee and was unanimously awarded with honors (*cum laude*), which is conferred to less than 5 % of Ph.D. candidates of Delft University of Technology. The dissertation has also won the *2014 Victor Lesser Distinguished Dissertation Runner-up Award* in recognition of an exceptional and highly impressive Ph.D. dissertation in the area of autonomous agents and multiagent systems. I am proud to now see it appear in *Springer Theses*, recognizing its excellence and high impact on research.

During his Ph.D., Tim Baarslag has successfully cooperated with leading researchers in the field, including Prof. Nicholas R. Jennings and Prof. Sarit Kraus. A prime example is the joint publication in *artificial intelligence* about the international negotiation competition he coorganized. The journal editor, G. Lakemeyer, commented as follows on this paper: "*(…) the paper is an excellent example of the kinds of reports we would like to see published (…), providing deep insights into the state of the art of a particular kind of AI systems, from which lessons for future developments can be learned.*"

All research results presented in this dissertation are published in top journals in the field; e.g., *artificial intelligence, decision support systems, autonomous agents and multiagent systems, and computational intelligence*. His work on the significance of automated negotiators provides a new perspective on the challenges in the field and resulted in a publications in leading conferences on AI, such as the *12th International Conference on Autonomous Agents and Multi-agent Systems (AAMAS 2013)* and the *21st European Conference on Artificial Intelligence (ECAI 2014)*. The success of his ideas is demonstrated throughout this book; for example, his results on negotiation learning techniques won the Best Paper Award in the *IEEE/WIC/ACM International Joint Conferences on Web Intelligence (WI) and Intelligent Agent Technologies (IAT)*.

To top it off, an agent that successfully applied his dissertation framework won first prize in *The 2013 International Automated Negotiating Agents Competition (ANAC)*.

Delft

December 2015

Prof. Catholijn M. Jonker

Dr. Koen V. Hindriks

Acknowledgements

I have thoroughly enjoyed my life as a Ph.D., and I could not have done it without the help and support—and the love and warmth—of the people around me.

First and foremost, I would like to thank my copromotor Koen Hindriks and my promotor Catholijn Jonker for all their suggestions and useful comments. Koen, thank you for being so dedicated and involved in my work, while at the same time giving me the freedom to pursue my own research ideas. Thank you for all the help, for having a supervision style based on trust, and for putting your faith in my proposals.

Catholijn, thank you for believing in me and for supporting me when it mattered most. Thank you for the intense discussions, for the brainstorm sessions, and for the opportunity to go to *AAMAS 2010*, just a few months into my Ph.D. It gave me a lot of inspiration, and it really kick-started my research. Thank you for the nice words and encouragements; these little things matter a lot.

Of course, I would also like to thank my colleagues of the *Interactive Intelligence Group* for making these years so much more enjoyable. Many of you have undoubtedly skipped straight to this section, but I like to think you are just saving the rest for later. In particular, I thank Hani Alers for his helpful advice and for organizing all kinds of cool stuff; Reyhan Aydogan for her cheerfulness and for being my surrogate mom in Delft; Joost Broekens for being so passionate and positive; Christian Detweiler for the music, the stage-diving, and the useless trivia; Nike Gunawan for her good mood and organizing the awesome Antwerp trip; Maaike Harbers for the "voeten op tafel" conversations about life and career; Alex Kayal for bringing a much-needed new impulse to the group, for the conversation slipups, and for making Minnesota awesome; Iris van de Kieft for sharing the experience of embarking on our Ph.D. adventure; Thomas King for his wits and for his innovative guitar techniques; Iulia Lefter for patiently answering all my questions about defense regulations; Hantao Liu for all the postdoc advice; Arman Noroozian for the math and puzzles and for his strength; Alina Pommeranz for the hallway laughs and for introducing me to Frisbee; Judith Redi for a healthy dose of realism; Dmytro Tykhonov for his wise tips and advice; Wietske Visser for her kindness and helpfulness; and Chang Wang for the history lessons and the impeccable impersonations.

Furthermore, I thank Anita Hoogmoed, Ruud de Jong, and Bart Vastenhouw for their unwavering support and cheerfulness and for always being willing to lend a hand.

I also received great support from outside the *TU Delft*. I thank my coauthors and collaborators from *The University of Southampton*, in particular the *Agents, Interaction and Complexity Group*, with special thanks to Nicholas Jennings, Enrico Gerding, Valentin Robu, and Colin Williams.

Great thanks also go to my colleagues at *Bar-Ilan University* and *Ben-Gurion University of the Negev*, especially Sarit Kraus and Raz Lin for all the collaborative and organizational effort and Kobi Gal for all the support. I thank Pınar Yolum from *Boğaziçi University* for our collaborations together, and a special thanks goes to Ivan Marsa-Maestre for helping me turn the page in Taipei.

I thank my sensei away from home, Takayuki Ito, for his help and for generously hosting me at *Nagoya Institute of Technology*. I also enjoyed collaborating with Katsuhide Fujita, now at *Tokyo University of Agriculture and Technology*. I thank Shantanu Chakraborty, Rafik Hadfi, and Shun Okuhara for making my stay at Nagoya such a blast. I am also grateful for a great and inspiring time in *Shizuoka University*, thanks to the warm welcome of Naoki Fukuta and Yoshinori Tsuruhashi.

A big thank you goes to my master students, Alex Dirkzwager and Mark Hendrikx, who I supervised with pleasure for the last two years. Without you, I would not have been able to do all the work I wanted to do. You both did an amazing job.

I also like to thank my housemates of the Jesseplaats for dropping by on all sorts of occasions, preferably uninvited, and for making sure I never came home to an empty house after a conference. I thank my band members and good friends Ivo Esseveld and Evgeny Rezunenko for making my life more musical and for occasionally showing up for band practice. I would like to thank and remember Mississippi John Hurt, John Fahey, and Jack Rose for making me want to pick up the guitar every day.

Derk van Veen and Tamara Vreeburg, thank you for being there for me through the good and the bad times. Thank you Elwin Man, for more than 15 years of beers, music, and countless chess blitz matches. I thank my trustworthy paranymphs Andreas Goetze and Hugo Looijestijn for more than a decade of yearly holidays, mathematics, and endless bar talks.

I thank my parents for everything. Thank you Ed, for giving me love for science. It makes me sad that you will not see how helpful and inspiring our talks have been for me, but you will always be there in my heart. Thank you so much Willy, for your love and encouragement, and for being such an awesome mom. I love you both.

Lastly, I thank my sweet Christina for her love, and for accepting me; for broadening my perspective on life; and for adding a welcome dash of Mediterranean spirit to my life in Delft.

Contents

About the Author

Tim Baarslag was born on May 27, 1983, in Utrecht, The Netherlands. He studied mathematics at the Faculty of Mathematics and Computer Science of Utrecht University. Tim received his propaedeutic degree in mathematics and computer science in 2012 (with honors) on the topic of the foundations of mathematics. Tim obtained his master's degree in mathematics in 2007 (with honors), with a focus on the complexity analysis of recursive algorithms. Tim obtained an additional bachelor's degree in computer science at Utrecht University in 2008 (with honors). He completed his master of science teaching mathematics in the Graduate School of Teaching and Learning, University of Amsterdam in 2009.

In 2014, Tim obtained his Ph.D. (with honors) at the Faculty of Electrical Engineering, Mathematics and Computer Science of Delft University of Technology on the topic of intelligent decision support systems for automated negotiation. During this period, Tim has twice been a visiting researcher at the Nagoya Institute of Technology of Japan.

As of June 2014, Tim is a research fellow in the Agents, Interaction and Complexity Group at the University of Southampton, where he works on devising intelligent negotiation systems that help users obtain better protection of their privacy-sensitive data.

Chapter 1
Introduction

Leo Hendrik Baekeland, born in 1863, was always at the head of his class. He graduated at sixteen and received his doctor's degree maxima cum laude when he was still only twenty-one. By 1891, he had opened an office in the US as an independent consultant and invented a type of photographic paper that could be developed under artificial light. In 1899, Leo Baekeland was still struggling with his Velox photosensitive manufacturing business. One day he received an invitation letter from George Eastman-Kodak, who had established the Eastman Kodak Company in Rochester, New York. George suggested that if Baekeland was willing to sell his Velox manufacturing company, he was welcome to visit him for a talk. During the long carriage ride up to Rochester, Baekeland planned to ask for $50,000, but kept wondering if he would be able to get even $25,000 for his manufacturing process. George Eastman invited Leo Baekeland into his office, and fortunately for Baekeland, Eastman spoke first and right away offered him one million dollars. Baekeland immediately took the offer. He could now afford to do his research in a well-equipped laboratory and went on to invent the first plastic, Bakelite.

1.1 Negotiation

Negotiation is a core activity in human society to form alliances, to reach trade agreements, and to resolve conflicts. One cannot overstate the importance of negotiation and the centrality it has taken in our everyday lives. People negotiate everywhere, in business as well as their personal lives [1], mostly without realizing they do so [2]. Negotiation not only occurs in obvious instances, such as job negotiation, politics, acquiring a house, or haggling at the marketplace. We also use it in various everyday situations, such as setting a calendar date with a friend, asking for a refund, or agreeing on a deadline.

The field of negotiation is an important topic of research in economics [3, 4], artificial intelligence [5–10], game theory [3, 5, 6, 9, 11–13], and social psychology [14]. The last two decades have seen a growing interest in the *automation* of negotiation and e-negotiation systems [6, 8, 15–17], for example in the setting of e-commerce [18–21]. This interest is fueled by the promise of automated agents

© Springer International Publishing Switzerland 2016
T. Baarslag, *Exploring the Strategy Space of Negotiating Agents*,
Springer Theses, DOI 10.1007/978-3-319-28243-5_1

being able to negotiate on behalf of human negotiators, and to find better outcomes than human negotiators [18, 21–25].

Negotiation agents can alleviate some of the efforts required of people during negotiations and make negotiation problems more manageable and comprehensible for negotiators [26]. The potential benefits of automation include the reduced time and negotiation costs resulting from automation [21, 27–29], the potential increase in negotiation usage since the user can avoid social confrontation [21, 30], the ability to improve the negotiation skills of the user [24, 31, 32], and the possibility of finding more interesting deals by exploration of the outcome space [21, 31]. There are also many unexpected uses of automated negotiation; for example controlling the load in an electrical grid [33], locating available parking spaces [34], playing *Civilization IV* [35], routing telephone calls [2], or Mars rovers coordinating autonomously who is better equipped for a given task [36]. Thus, success in developing an automated agent with negotiation capabilities has great advantages and implications.

1.2   An Automated Negotiator

Automated negotiation research deals with two main topics [6, 37], which both have received their fair share of attention in the field.

From a system design or mechanism design point of view, devising an effective *negotiation protocol* is the most important concern (e.g. [2, 38–40]). Negotiation protocols are the set of rules that govern the way the negotiation takes place. This covers the number of participants and the valid actions of the participants in every particular negotiation state (e.g., which messages can be sent by whom, to whom, and at what stage). It also specifies the structure of the possible agreements, and what operations are allowed to change the contents of proposed offers.

In other cases, such as in this thesis, the *agent's decision making model* is the dominant concern (e.g. [8, 41, 42]). The main focus here is on the reasoning modules and strategies that the negotiating agents employ to make their decision in order to achieve their objectives. When the protocol is such that it leaves room for strategic reasoning, the success of a self-interested agent is determined by the effectiveness of its decision making model.

In order to be successful, a negotiating agent needs to be able to perform a variety of tasks. First of all, the agent needs to be able to interact with the others in a given *negotiation setting* that defines the different parameters of the negotiation (see Fig. 1.1). During negotiation, the agent exchanges proposals with the other participants in order to reach an acceptable agreement, which is a contract that all negotiating parties agree upon. The range of contracts being negotiated over (i.e., the set of all possible negotiation outcomes) is called the *negotiation domain*. Of course, the proposals must be submitted according to certain rules and be valid according to constraints set by the negotiation protocol. Every agent has *preferences* over the negotiation domain, which define the particular *negotiation scenario*.

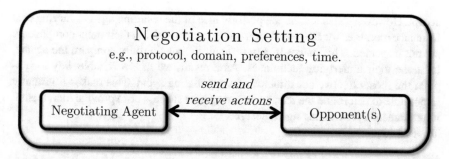

Fig. 1.1 The setting for an automated negotiator

The agent designer can select a number of *performance measures* to assess the success of a negotiating agent. The most popular way is to assign a certain *utility* to the outcomes that are reached by the agent. Other measures that the agent designer might choose are the duration of the negotiation (i.e., how fast the agent is able to reach agreements), or fairness of the outcome (i.e., whether the agreement satisfies all negotiation parties).

1.2.1 Generic Negotiation Strategies

With the constant introduction of new negotiation domains, negotiating agents may encounter different types of opponents with different characteristics. Therefore, an important research topic in automated negotiation is the design of agents that can perform well in a *variety of circumstances*. Such generic automated negotiation agents should be capable of negotiating proficiently within arbitrary negotiation scenarios, with opponents that are diverse in their behavior.

A number of automated negotiation strategies have been proposed that are designed to operate in specific and relatively simple scenarios and are often based on simplifying assumptions (e.g., [32, 39, 43–46]). A typical example of such an assumption is that the opponent strategies and preferences are known or partially known. This is generally unrealistic, as negotiators tend to avoid revealing their private information [47], because the shared information may be used to the revealer's disadvantage [48].

Examples of more general agent negotiators are increasingly available in the literature. Every year, automated negotiation agents are improving in various ways and have proven to be successful in many regards (for an exposition, see Chap. 2 and Appendix B). They all have their unique strengths and weaknesses and are based on a variety of techniques, such as game-trees [49], generic trade-off algorithms [43, 45], concession curves [50, 51], statistical analysis [52, 53], wavelet decomposition [54–56], and Gaussian process regression [57–59].

Each technique is used for various aspects of the negotiation process, such as preference learning, strategy prediction, making concessions, or choosing when to accept. However, when testing the performance of the resulting agents, varying performance measures for the negotiation outcome are used, but their inner components are not inspected. This makes it very difficult to meaningfully compare the agents, let alone their underlying techniques. As a result, we lack a reliable way to pinpoint the most effective constituents of a negotiating agent. This makes it virtually impossible to determine the reasons for an agent's success or to provide incremental improvement over existing agent designs.

To put it succinctly:

> **Problem** We lack a fundamental approach to build comparably effective, general automated negotiators in an incremental fashion that enables us to understand how their underlying techniques influence their performance.

1.3 Bidding, Learning, and Accepting

There is a wide variety of currently existing sophisticated agent strategies and architectures, but we show in this thesis that there is some common structure to their overall design. For example, every agent decides whether the opponent's offer is acceptable, and if not, what offer should be proposed instead. In addition, when the agent decides on the counter-offer, it considers its own utility, but it usually also takes the opponent's utility into account. We elaborate on this topic in Chap. 3, but for now it suffices to say that we distinguish three distinct *components* of a negotiating agent strategy, each of which we analyze separately in this thesis:

- **Bidding strategy**. Given the current negotiation state, what are the appropriate bids to be made?
- **Opponent model**. How can we learn what the opponent wants, and how do we take this into account?
- **Acceptance strategy**. Should we accept the opponent's bid, reject it, or walk away from the negotiation altogether?

There are two major advantages of distinguishing between the different components of a negotiating agent's strategy: first, given performance measures for the individual components, it allows the study of the behavior and performance of the components in isolation. For example, it becomes possible to compare the accuracy of the opponent modeling components of a set of agents, and to pinpoint the best opponent model among them. Second, we can assemble, from already existing components, new negotiating agents in a plug and play fashion (see Fig. 1.2), e.g.: replacing the opponent model of an agent and then examining whether this makes a difference in performance. Such a procedure enables us to combine the individual components to systematically explore the space of possible negotiation strategies. Finding a good negotiating strategy then boils down to deciding *what to bid*, *how to learn*, and *when to accept*.

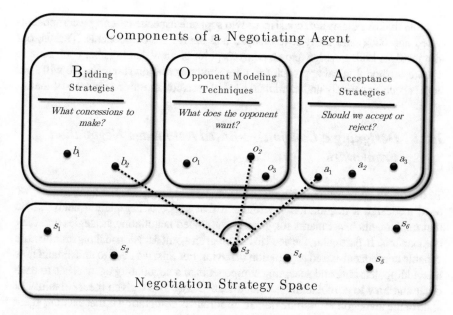

Fig. 1.2 The strategy space of automated negotiators can be explored by combining a bidding strategy with an opponent model and an acceptance strategy

Due to possible dependencies between the components, the agent should be able to combine them in a meaningful way; e.g., purposely selecting 'exploratory offers' to learn more about the opponent's preferences, or considering the opponent's future behavior when deciding whether to accept the opponent's bid. This means that in order to be successful, a negotiating agent should not only have the three components work effectively in an individual manner, but the agent also needs a powerful architecture with which to assemble the components into a negotiation strategy.

1.4 Research Questions

The advantages of a component-based approach for an automated negotiator as outlined above have motivated our concrete research aim as follows:

Thesis Aim

The central aim of this thesis is to research effective ways for a general automated negotiating strategy *to learn*, *to bid*, and *to accept* and to develop a compositional approach for evaluating and combining these components.

Note that our thesis aim consists of two separate aspects: *creating* a component-based approach, and *using* it to analyze and devise the components. That is, our aim involves both a *design* and an *analysis* point of view that together contribute to a more methodological approach for automated negotiation research. We will treat both aspects separately and formulate a set of research questions for each of them.

1.4.1 Designing a Component-Based Automated Negotiation Framework

To develop a compositional approach to evaluate and combine the components, we need to design a negotiation environment that supports negotiation analysis and that implements benchmarks for general automated negotiating strategies and their components. In particular, we need to establish an agent decision making architecture capable component-based negotiation behavior. For this, we need to understand how the bidding, learning and accepting components of a negotiating agent relate to each other and how to combine them in an effective way. Also, given the availability of state of the art negotiation strategies, an important consideration is that existing agent designs can be incorporated into our approach.

Thus, in order to achieve the design aspect of our aim, we address the following questions:

Research Questions I
Designing a component-based automated negotiation framework.

How do we create a negotiation framework that:

1. supports new agent designs and provides insight into the effectiveness of negotiation strategies;
2. facilitates evaluating and combining various negotiation strategy components;
3. enables us to decompose existing, state of the art agent designs into distinct components.

1.4.2 Analyzing the Negotiating Strategy Components

To analyze the components individually, it is necessary to formulate benchmarks and predictors for the performance of the individual components. The performance measures for the bidding strategy, opponent model, and acceptance strategy are likely to be different for each case. With performance measures for every component, we can specify solutions separately in a plug and play fashion. We will consider specific situations (and specific classes of opponents in particular) for which we can find effective solutions, and in some cases, even optimal ones.

Of course, after analyzing the components individually, we need to consider what happens if we assemble them again, and whether combining effective components also improves the overall performance. There could be strong interdependency between the components, and some components can prove to be more important to consider than others.

We formulate three additional questions regarding the analysis aspect of our research aim:

Research Questions II
Analyzing the negotiating strategy components: what to bid, how to learn, and when to accept.

1. What measures can we use to compare and predict the performance of the individual components?
2. Can we pinpoint classes of opponents against which we can find effective components? Can we formulate optimal solutions for any of the components?
3. How does the performance of the components influence the negotiator's performance as a whole, and which components are most important?

1.5 Thesis Scope

Before we describe our research method to answer our research questions, we briefly frame the scope of our work. We will elaborate extensively on our model of negotiation (and on related possibilities) in Chap. 2.

This thesis focuses on bilateral negotiations (i.e., negotiations between two agents), in which the agents exchange offers in turns. While the negotiation domain is known by both agents, the preferences of each player is private information. The agents seek to reach an agreement while aiming to satisfy their own preferences.

The heart of this thesis consists of the analysis of decision making procedures for a negotiating agent in such a setting. More specifically, if we adhere to the classification used by Lomuscio et al. [21], the focus of this thesis is as follows:

Thesis Scope
This thesis focuses on *one-to-one* negotiations with *alternating offers* on *multiple-issue* domains, using *self-interested* agents with *bounded rationality* and *incomplete information*.

1.6 Dissertation Outline

We give a detailed overview of this thesis in the paragraphs below (see also Fig. 1.3).
A summary is available at the end of this section.

Quick Read Guide
For the reader in a hurry, we suggest the following quick read guide for this
thesis:

1. Skip the Background chapter entirely, but read the summaries of Appendix
   A about GENIUS (p. 215) and Appendix B about ANAC (p. 223).
2. Read the full chapter about the BOA framework (Chap. 3, p. 53).
3. Choose one chapter for each component of the BOA framework. We recommend Chap. 5 on optimal acceptance policies (p. 91), Chap. 7 on performance and accuracy of learning methods (p. 129), and Chap. 8 on the
   classification of bidding strategies (p. 147).
4. Read Sect. 10.4 (p. 186) on how the BOA components fit together.
5. End with our concluding chapter (p. 195).

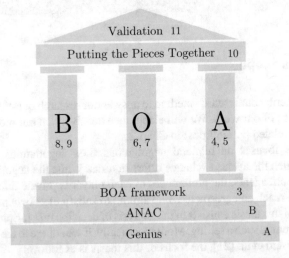

Fig. 1.3 A graphical representation of the outline of this thesis. GENIUS (Appendix A) lays the
groundwork for ANAC (Appendix B), and the BOA framework (Chap. 3) builds on top of both
of them. In turn, all three support the pillars to component analysis of *Bidding* (Chaps. 8 and 9),
Opponent modeling (Chaps. 6 and 7), and *Accepting* (Chaps. 4 and 5). We put the pieces together
in Chap. 10, culminating in the validation of the BOA framework (Chap. 11)

1.6.1 The Fundamentals

We start Chap. 2 by briefly discussing the background and related work in automated negotiation. We give definitions of the basic terminology used in negotiation literature and we discuss prime examples of existing automated negotiation architectures and strategies. We focus specifically on existing bidding strategies, opponent models and acceptance strategies, and on combining a set of components to explore the negotiation strategy space.

We conclude the background chapter by describing several methodologies for evaluating and comparing negotiation strategies and components. Among our discussed evaluation methods are performance and accuracy measures, agent competitions, and analytical software to assess the outcome of the negotiation. We conclude with a discussion of several evaluation methodologies of negotiation strategies, with an emphasis on *performance* and *accuracy* measures.

1.6.2 The BOA Architecture

Chapter 3 describes the BOA architecture, in which we can develop and integrate the different components of a negotiating agent into one negotiating strategy. We use the BOA architecture to explore the space of possible strategies by studying and recombining different state of the art strategy components.

The BOA architecture is integrated seamlessly into a generic negotiation environment called GENIUS (Appendix A), which is a flexible software environment that facilitates the design, evaluation and analysis of negotiation strategies. GENIUS provides full support for a diversity of different negotiation protocols, scenarios, and agents, which we amend with analytical tools and various existing agents, negotiation scenarios, and protocols from literature. The implementation of the BOA architecture offers the user the ability to create and combine newly developed components using a graphical user interface.

To explore the negotiation strategy space of the negotiation research community, we require a variety of different state of the art negotiating agents, and we need to formulate objective evaluation criteria for them. Appendix B describes the organization and insights gained from four instances of a yearly international negotiation competition (ANAC) held between 2010 and 2013 in conjunction with the International Conference on Autonomous Agents and Multiagent Systems (AAMAS). ANAC acts as an evaluation tool for negotiation strategies, and encourages the design of negotiation strategies and scenarios. Moreover, through ANAC we learn new, improved approaches to effective agent designs, which are accessible as benchmarks for the negotiation research community. We organize the competition, but we also *participate* in it, through which we foster our ties with the automated negotiation community. The agents, domains, and scores of ANAC are used in most chapters of this thesis and are discussed in detail in Appendix C–F.

With GENIUS, ANAC and the BOA architecture in place, we embark on the task to re-implement more than 20 agents from literature and ANAC and to *decouple* them to fit into the BOA architecture without introducing any changes in their behavior. This enables us to do two things: first of all, it allows us to independently analyze the components of every decoupled negotiation strategy; second, we can proceed to *mix and match* different BOA components to create new negotiation strategies. Such a procedure makes it possible to systematically search for an effective automated negotiator.

1.6.3 Analyzing the Components of an Automated Negotiator

In Chaps. 4–9, we focus on the first benefit of the BOA architecture: seeking out the best of each BOA component. For each of the three components, we find and analyze the best ones for specific cases, and in the case of bidding and acceptance strategies, we devise optimal ways of doing so.

In Chap. 4, we study and classify all current approaches regarding *acceptance strategies*, and we pinpoint the ones that perform best, together with reasons why they work well. In Chap. 5, we adopt a more principled approach by applying optimal stopping theory to calculate the optimal decision on the acceptance of an offer.

We study the performance of a variety of different *opponent models* in Chap. 6, identifying the best preference learning techniques. We consider opponent models from both a performance and an accuracy perspective in Chap. 7, and we pinpoint the accuracy measures that are the best *predictors* for good performance of opponent modeling techniques.

Finally, we take two different approaches to gain more insight into effective *bidding strategies*. In Chap. 8, we present a new classification method for negotiation strategies, based on their pattern of concession making, and we formulate guidelines on how agents should bid in order to be successful. We focus on *optimal* bidding strategies in Chap. 9. We apply optimal stopping theory again, this time to find concession sequences that maximize the utility for the bidder against particular opponents. We show there is an interesting connection between optimal bidding and optimal acceptance strategies, in the sense that they are mirrored versions of each other.

1.6.4 Putting the Pieces Together

Lastly, after analyzing all components separately, we put the pieces back together again in Chap. 10, showing that the BOA framework leads to significant improvements in agent design. We win ANAC 2013, which had 19 participating teams from 8 international institutions, with an agent that is designed using the BOA framework and is informed by our analysis of the different components. We take all BOA components accumulated so far, including the best ones, and combine them all together to explore the space of negotiation strategies. We test the performance of every component and perform statistical analysis to see whether the best components together lead to the best agent, and which components contribute most significantly to the end result.

Dissertation outline summary

1. Design a component-based negotiation architecture (BOA);
2. Establish a negotiation environment (GENIUS) and integrate the BOA architecture in it;
3. Organize a negotiation competition (ANAC);
4. Fit existing, state of the art negotiating agents (including ANAC agents) into the BOA framework;
5. Analyze and optimize all BOA components independently;
6. Recombine the BOA components, evaluate the component contributions, and benchmark the resulting agents.
7. Validate the BOA framework and demonstrate its value by winning ANAC 2013 with a BOA agent.

1.7 Contributions

The most important contributions of this thesis are listed below. We elaborate on each contribution in our conclusions in Chap. 11.

Contributions

1. Introducing a component-based negotiation architecture to systematically explore the space of automated negotiation strategies (Chap. 3).
2. Developing design, evaluation and benchmarking methods for negotiation agents (Appendix A and B).
3. Classifying and comparing acceptance strategies and formulating optimal acceptance strategies (Chaps. 4 and 5).
4. Identifying the most effective and accurate learning methods, and determining the best methods to predict their performance (Chaps. 6 and 7).
5. Formulating optimal bidding strategies and categorizing concession behavior (Chaps. 8 and 9).
6. Quantifying the importance and interactions of the components of a negotiating agent (Chap. 10).
7. Validating the BOA architecture and demonstrating its success in exploring the negotiation space (Chaps. 10 and 11).

References

1. Raiffa H, Richardson J, Metcalfe D (2003) Negotiation analysis: the science and art of collaborative decision making. Harvard University Press, Cambridge
2. Rosenschein JS, Zlotkin G (1994) Rules of encounter: designing conventions for automated negotiation among computers. MIT Press, Cambridge
3. Osborne MJ, Rubinstein A (1990) Bargaining and markets (Economic theory, econometrics, and mathematical economics). Academic Press, London
4. Raiffa H (1982) The art and science of negotiation: How to resolve conflicts and get the best out of bargaining. Harvard University Press, Cambridge
5. Gerding EH, Bragt DDB, La Poutré JA (2000) Scientific approaches and techniques for negotiation: a game theoretic and artificial intelligence perspective. Technical report, CWI (Centre for Mathematics and Computer Science), Amsterdam, The Netherlands
6. Jennings NR, Faratin P, Lomuscio AR, Parsons S, Wooldridge MJ, Sierra C (2001) Automated negotiation: prospects, methods and challenges. Group Decis Negot 10(2):199–215
7. Kraus S (1997) Negotiation and cooperation in multi-agent environments. Artif Intell 94(1–2):79–97
8. Kraus S (2001) Strategic negotiation in multiagent environments. MIT Press, Camcridge
9. Li C, Giampapa J, Sycara KP (2003) A review of research literature on bilateral negotiations. Technical report, Robotics Institute, Pittsburgh, PA
10. Silaghi GC, Şerban LD, Litan CM (2010) A framework for building intelligent SLA negotiation strategies under time constraints. In: Altmann J, Rana OF (eds) Proceedings of economics of grids, clouds, systems, and services: 7th international workshop, vol 6296. Springer, New York, p 48
11. Binmore K, Vulkan N (1999) Applying game theory to automated negotiation. Netnomics 1(1):1–9
12. Liang Y, Yuan Y (2008) Co-evolutionary stability in the alternating-offer negotiation. In: IEEE Conference on cybernetics and intelligent systems, pp 1176–1180
13. Rubinstein A (1982) Perfect equilibrium in a bargaining model. Econometrica 50(1):97–109
14. Rubin JZ, Brown BR (1975) The social psychology of bargaining and negotiation. Academic press, New York
15. Beam C, Segev A (1997) Automated negotiations: a survey of the state of the art. Wirtschaftsinformatik 39(3):263–268
16. Guttman RH, Maes P (1999) Agent-mediated integrative negotiation for retail electronic commerce. In: Noriega P, Sierra C (eds) Agent mediated electronic commerce. Lecture notes in computer science, vol 1571, Springer, Berlin, pp 70–90
17. Kersten GE, Lai H (2007) Negotiation support and e-negotiation systems: an overview. Group Decis Negot 16(6):553–586
18. Bosse T, Jonker CM (2005) Human vs. computer behaviour in multi-issue negotiation. In: Proceedings of the rational, robust, and secure negotiation mechanisms in multi-agent systems, RRS '05. IEEE Computer Society, Washington, DC, USA, pp 11–24
19. He M, Jennings NR, Leung H (2003) On agent-mediated electronic commerce. IEEE Trans Knowl Data Eng 15(4):985–1003
20. Kowalczyk R, Ulieru M, Unland R (2003) Integrating mobile and intelligent agents in advanced e-commerce: a survey. In: Carbonell JG, Siekmann J, Kowalczyk R, Müller JP, Tianfield H, Unland R (eds) Agent technologies, infrastructures, tools, and applications for E-Services. Lecture notes in computer science, vol 2592. Springer, Berlin, pp 295–313
21. Lomuscio AR, Wooldridge MJ, Jennings NR (2003) A classification scheme for negotiation in electronic commerce. Group Decis Negot 12(1):31–56
22. Dzeng R-J, Lin Y-C (2005) Searching for better negotiation agreement based on genetic algorithm. Comput Aided Civil Infrastruct Eng 20(4):280–293
23. Jazayeriy H, Azmi-Murad M, Sulaiman MN, Udzir NI (2011) A review on soft computing techniques in automated negotiation. Sci Res Essays 6(24):5100–5106

24. Lin R, Kraus S (2010) Can automated agents proficiently negotiate with humans? Commun ACM 53(1):78–88
25. Oshrat Y, Lin R, Kraus S (2009) Facing the challenge of human-agent negotiations via effective general opponent modeling. In: Proceedings of the 8th international conference on autonomous agents and multiagent systems, AAMAS '09. International foundation for autonomous agents and multiagent systems, vol 1. Richland, SC, pp 377–384
26. Delaney MM, Foroughi A, Perkins WC (1997) An empirical study of the efficacy of a computerized negotiation support system (NSS). Decis Support Syst 20(3):185–197
27. Bui HH, Venkatesh S, Kieronska DH (1995)An architecture for negotiating agents that learn. Technical report, Department of Computer Science, Curtin University of Technology, Perth, Australia
28. Bui HH, Venkatesh S, Kieronska DH (1999) Learning other agents' preferences in multi-agent negotiation using the bayesian classifier. Int J Coop Inf Syst 8(4):273–293
29. Carbonneau RA, Kersten GE, Vahidov RM (2008) Predicting opponent's moves in electronic negotiations using neural networks. Expert Syst Appl 34(2):1266–1273
30. Broekens J, Jonker CM, Meyer JJC (2010) Affective negotiation support systems. J Ambient Intell Smart Environ 2(2):121–144
31. Hindriks KV, Jonker CM (2009) Creating human-machine synergy in negotiation support systems: towards the pocket negotiator. In: Proceedings of the 1st international working conference on human factors and computational models in negotiation, HuCom '08. ACM, New York, pp 47–54
32. Lin R, Oshrat Y, Kraus S (2009) Investigating the benefits of automated negotiations in enhancing people's negotiation skills. In: AAMAS '09: proceedings of the 8th international conference on autonomous agents and multiagent systems. International foundation for autonomous agents and multiagent systems. Richland, SC pp 345–352
33. Jennings NR, Mamdani EH (1992) Using joint responsibility to coordinate collaborative problem solving in dynamic environments. In: 10th National conference on artificial intelligence (AAAI-92), pp 269–275
34. Di Napoli C, Di Nocera D, Rossi S (2014) Negotiating parking spaces in smart cities. In: Proceeding of the 8th international workshop on agents in traffic and transportation, in conjunction with AAMAS
35. Afiouni EN, Ovrelid LJ (2013) Negotiation for strategic video games. Master's thesis, Norwegian University of Science and Technology, Department of Computer and Information Science
36. Clement BJ, Barrett AC (2003) Continual coordination through shared activities. In: Proceedings of the second international joint conference on Autonomous agents and multiagent systems, AAMAS '03. ACM, New York, pp 57–64
37. Kraus S (2001) Automated negotiation and decision making in multiagent environments. Multi-Agent Syst Appl 1:150
38. Fatima SS, Wooldridge MJ, Jennings NR (2007) On efficient procedures for multi-issue negotiation. Agent-Mediated Electron Commer 4452:31–45 249
39. Ito T, Hattori H, Klein M (2007) Multi-issue negotiation protocol for agents: exploring nonlinear utility spaces. In: Proceedings of the 20th international joint conference on artifical intelligence, IJCAI'07. Morgan Kaufmann Publishers Inc, San Francisco, pp 1347–1352
40. Vulkan N, Jennings NR (2000) Efficient mechanisms for the supply of services in multi-agent environments. Decis Support Syst 28(1–2):5–19
41. Sandholm T, Vulkan N (1999) Bargaining with deadlines. In: Proceedings of the sixteenth national conference on Artificial intelligence and the eleventh Innovative applications of artificial intelligence conference, AAAI '99/IAAI '99. American Association for Artificial Intelligence. Menlo Park, pp 44–51,
42. Sierra C, Faratin P, Jennings NR (1997) A service-oriented negotiation model between autonomous agents. In: Boman M, van de Velde W (eds) Proceedings of the 8th European workshop on modelling autonomous agents in multi-agent world, MAAMAW-97. Lecture notes in artificial intelligence, vol 1237. Springer, Berlin, pp 17–35

43. Faratin P, Sierra C, Jennings NR (2002) Using similarity criteria to make issue trade-offs in automated negotiations. Artif Intell 142(2):205–237
44. Fatima SS, Wooldridge MJ, Jennings NR (2005) A comparative study of game theoretic and evolutionary models of bargaining for software agents. Artif Intell Rev 23(2):187–205
45. Jonker CM, Robu V, Treur J (2007) An agent architecture for multi-attribute negotiation using incomplete preference information. Auton Agents Multi-Agent Syst 15:221–252
46. Kraus S, Hoz-Weiss P, Wilkenfeld J, Andersen DR, Pate A (2008) Resolving crises through automated bilateral negotiations. Artif Intell 172(1):1–18
47. Boles TL, Croson RTA, Murnighan JK (2000) Deception and retribution in repeated ultimatum bargaining. Organ Behav Hum Decis Process 83(2):235–259
48. Murnighan JK, Babcock L, Thompson L, Pillutla M (1999) The information dilemma in negotiations: effects of experience, incentives, and integrative potential. Int J Confl Manage 10(4):313–339
49. Karp AH, Wu R, Chen KY, Zhang A (2004) A game tree strategy for automated negotiation. In: Proceedings of the 5th ACM conference on Electronic commerce, EC '04. ACM, New York, pp 228–229
50. Faratin P, Sierra C, Jennings NR (1998) Negotiation decision functions for autonomous agents. Robot Auton Syst 24(3–4):159–182
51. Fatima SS, Wooldridge MJ, Jennings NR (2002) Multi-issue negotiation under time constraints. In: AAMAS '02: proceedings of the first international joint conference on Autonomous agents and multiagent systems. ACM, New York, pp 143–150
52. Kawaguchi S, Fujita K, Ito T (2012) Compromising strategy based on estimated maximum utility for automated negotiating agents. In: Ito T, Zhang M, Robu V, Fatima S, Matsuo T (eds) New trends in agent-based complex automated negotiations. Series of studies in computational intelligence. Springer, Berlin, pp 137–144
53. Kawaguchi S, Fujita K, Ito T (2013) AgentK2: compromising strategy based on estimated maximum utility for automated negotiating agents. In: Ito T, Zhang M, Robu V, Matsuo T (eds) Complex automated negotiations: theories, models, and software competitions. Studies in computational intelligence, vol 435. Springer, Berlin, pp 235–241
54. Chen S, Weiss G (2012) An efficient and adaptive approach to negotiation in complex environments. In: De Raedt L, Bessiere C, Dubois D, Doherty P, Frasconi P, Heintz F, Lucas PJF (eds) ECAI. Frontiers in artificial intelligence and applications, vol 242. IOS Press, Amsterdam, pp 228–233
55. Chen S, Weiss G (2013) An efficient automated negotiation strategy for complex environments. Eng Appl Artif Intell 26:2613–2623
56. Chen S, Ammar HB, Tuyls K, Weiss G (2013) Optimizing complex automated negotiation using sparse pseudo-input gaussian processes. In: Proceedings of the 2013 international conference on autonomous agents and multi-agent systems, AAMAS '13. International foundation for autonomous agents and multiagent systems. Richland, SC, pp 707–714
57. Williams CR (2012) Practical strategies for agent-based negotiation in complex environments. Ph.D. thesis, University of Southampton
58. Williams CR, Robu V, Gerding EH, Jennings NR (2012) Iamhaggler: a negotiation agent for complex environments. In: Ito T, Zhang M, Robu V, Fatima S, Matsuo T (eds) New trends in agent-based complex automated negotiations. Studies in computational intelligence. Springer, Berlin, pp 151–158
59. Williams CR, Robu V, Gerding EH, Jennings NR (2013) Iamhaggler 2011: a gaussian process regression based negotiation agent. In: Ito T, Zhang M, Robu V, Matsuo T (eds) Complex automated negotiations: theories, models, and software competitions. Studies in computational intelligence, vol 435. Springer, Berlin, pp 209–212

Chapter 2
Background

Abstract In this chapter, we discuss briefly the background and related work in automated negotiation. We begin with definitions of the key aspects of automated negotiation, such as the negotiation *domain*, the *protocol*, and the *preferences*. We discuss what it means for a negotiating agent to employ a *negotiation strategy* and we highlight several prime examples of existing negotiation strategies. We also discuss a number of high-level *negotiation architectures* and how they can assist in exploring the negotiation strategy space. We focus specifically on the three components we distinguish in the Chap. 1, namely the various ways in which current negotiation strategies *bid*, *learn*, and *accept*. We conclude the background chapter by describing several methodologies for evaluating and comparing negotiation strategies and components. Among our discussed evaluation methods are *performance* and *accuracy* measures, agent competitions, and analytical software to assess the outcome of the negotiation.

2.1 Introduction

Negotiation is a common and important process for making decisions and resolving conflicts. People encounter negotiation situations everywhere, from specific situations such as job negotiations and hostage crises situations [1] to more general situations such as resource and task allocation mechanisms [2–4], conflict resolution mechanisms [5, 6], and decentralized information services [7, 8].

In recent years, the fact that negotiation covers many aspects of our lives has led to an increasing focus on the design of *automated* negotiators; i.e., autonomous agents capable of negotiating with other agents in a specific environment [7, 9]. This interest has been growing since the beginning of the 1980s with the work of early adopters such as Smith's Contract Net Protocol [4], Sycara's PERSUADER [10, 11], Robinson's OZ [12], and the work by Rosenschein [13] and Klein [14].

In this chapter, we discuss briefly the background and related work in automated negotiation. We will begin with definitions of the basic terminology used in this

© Springer International Publishing Switzerland 2016

T. Baarslag, *Exploring the Strategy Space of Negotiating Agents*,
Springer Theses, DOI 10.1007/978-3-319-28243-5_2

field in Sect. 2.2. In the subsequent Sect. 2.3, we discuss several prime examples of existing negotiation strategies and their architecture. In Sect. 2.4 we discuss several ways of evaluating negotiation strategies.

2.2   Terminology

The defining elements of a bilateral negotiation are depicted in Fig. 2.1. A bilateral automated negotiation concerns a negotiation between *two* agents, usually called *A* and *B*. The party that is negotiated with is also called the *partner* or *opponent*.

The *negotiation setting* consists of the *negotiation protocol*—the rules of encounter—, the negotiating agents, and the *negotiation scenario*. The negotiation takes place in a *negotiation domain*, which specifies all possible outcomes (the so-called *outcome space*). Furthermore, every agent in the scenario has a *preference profile*, which expresses the preference relations between the possible outcomes. Together, this defines the *negotiation scenario* that takes place between the agents. The negotiation scenario and protocol specify the possible *actions* an agent can perform in a given negotiation state.

2.2.1   Negotiation Domain

The *negotiation domain*—or *outcome space*—is denoted by Ω and defines the set of possible negotiation outcomes. The *domain size* is the number of possible outcomes $|\Omega|$. A negotiation domain consists of one or more *issues*, which are the main resources or considerations that need to be resolved through negotiation; for example, the price or the color of a car that is for sale. Issues are also sometimes referred

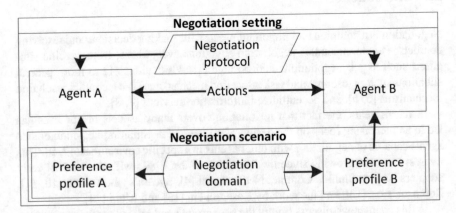

Fig. 2.1 Overview of the defining elements of an automated bilateral negotiation

to as attributes, but we reserve the latter term for *opponent attributes*, which are properties that may be useful to model to gain an advantage in a negotiation.

To reach an agreement, the agents must settle on a specific alternative or *value* for each negotiated issue. That is, an *agreement* on n issues is an outcome that is accepted by both parties of the form $\omega = \langle \omega_1, \ldots, \omega_n \rangle$, where ω_i denotes a value associated with the ith issue. We will focus mainly on settings with a finite set of discrete values per issue. A *partial agreement* is an agreement on a subset of the issues. We say that an outcome space defined by a single issue is a *single-issue negotiation*, and a *multi-issue negotiation* otherwise.

2.2.2 Negotiation Protocol

A negotiation protocol fixes the rules of encounter [15], specifying which actions each agent can perform at any given moment. Put another way, it specifies the admissible *negotiation moves*. There are a number of bilateral negotiation protocols. We do not aim to provide a complete overview of all protocols, instead we refer to Lomuscio et al. [16] for an overview of high-level parameters used to classify them, and to Marsa-Maestre et al. [17] for guidelines on how to choose the most appropriate protocol to a particular negotiation problem.

An often used negotiation protocol in bilateral automated negotiation is the *alternating offers protocol* [18, 19]. This protocol dictates that the two negotiating agents propose *outcomes*, also called *bids* or *offers*, in turns. That is, the agents create a *bidding history*: one agent proposes an offer, after which the other agent proposes a counter-offer, and this process is repeated until the negotiation is finished, for example by time running out, or by one of the parties accepting.

We use the alternating offers protocol throughout this thesis because of its simplicity, and moreover, it is a protocol which is widely studied and used in the literature, both in game-theoretic and heuristic settings (a non-exhaustive list includes [7, 18, 20–22]).

An important feature that differentiates protocols is their usage and definition of the *deadline* of a negotiation. The *deadline* of a negotiation refers to the time before which an agreement must be reached to achieve an outcome better than the best alternative to a negotiated agreement [23]. Each agent can have its own private deadline, or the deadline can be shared among the agents. The deadline may be specified as a maximum number of rounds [24], or alternatively as a real-time target. Note that when the negotiation happens in real time, the time required to reach an agreement depends on the deliberation time of the agents (i.e., the amount of computation required to evaluate an offer and produce a counter offer).

As in [25, 26], we supplement the alternating-offers protocol with a common global real time line, represented here by $\mathcal{T} = [0, D]$. We stipulate that the deadline has been reached when $t = D$, at which moment both agents receive utility 0.

We represent by $x^t_{A \rightarrow B}$ the negotiation outcome proposed by agent A to agent B at time t. A *negotiation thread* or *negotiation trace* (cf. [26, 27]) between two agents A and B at time $t \in \mathcal{T}$ is defined as a finite sequence

$$H^t_{A \leftrightarrow B} := \left(x^{t_1}_{p_1 \to p_2}, x^{t_2}_{p_2 \to p_3}, x^{t_3}_{p_3 \to p_4}, \dots, x^{t_n}_{p_n \to p_{n+1}} \right),$$

where

1. The offers are ordered over time $\mathcal{T}$: $t_k \leq t_l$ for $k \leq l$.
2. The offers are alternating between the agents: $p_k = p_{k+2} \in \{A, B\}$ for all k.
3. All t_i represent instances of time $\mathcal{T}$, with $t_n \leq t$,
4. The agents exchange complete offers: $x^{t_k}_{p_k \to p_{k+1}} \in \Omega$ for $k \in \{1, \dots, n\}$.

Additionally, the last element of $H^t_{A \leftrightarrow B}$ may be equal to one of the particles $\{Accept, End\}$. We will say a negotiation thread is *active* if this is not the case.

When agent A receives an offer $x^t_{B \to A}$ from agent B sent at time t, it has to decide at a later time $t' > t$ whether to accept the offer, or to send a counter-offer $x^{t'}_{A \to B}$. Given a negotiation thread $H^t_{A \leftrightarrow B}$ between agents A and B, we can express the action performed by A with a *decision function* [25, 26]. The resulting action is used to extend the current negotiation thread between the two agents. If the agent does not accept the current offer, and the deadline has not been reached, it will prepare a counter-offer by using a negotiation strategy or *tactic* to generate new values for the negotiable issues (see Sect. 2.3).

Various alternative versions of the alternating offers protocol have been used in automated negotiation, extending the default protocol, and imposing additional constraints; for example, in a variant called the *monotonic concession protocol* [15, 28], agents are required to initially disclose information about their preference order associated with each issue and the offers proposed by each agent must be a sequence of concessions, i.e.: each consecutive offer has less utility for the agent than the previous one. Other examples are the three protocols discussed by Fatima et al. [29] that differ in the way the issues are negotiated: simultaneously in bundles, in parallel but independently, and sequentially. The first alternative is shown to lead to the highest quality outcomes. A final example is relevant for our work in Chap. 9 on optimal concession curves, namely a protocol in which only *one* offer can be made. In such a situation, the negotiation can be seen as an instance of the ultimatum game, in which a player proposes a deal that the other player may only accept or refuse [30]. In [31], a similar bargaining model is explored as well; that is, models with one-sided incomplete information and one sided offers. It investigates the role of confrontation in negotiations and uses optimal stopping is to decide whether or not to invoke conflict. The setting of Chap. 9 can also be found in [32], which presents an alternating offer protocol for bilateral bargaining with imperfect information and deadline constraints.

2.2.3 Preference Profiles

Negotiating agents are assumed to have a *preference profile*, which is a *preference order* $\geq$ that ranks the outcomes in the outcome space. Preferences are said to be *ordinal* when they are fully specified by a preference order. Together with the domain they make up the *negotiation scenario*.

An outcome ω' is said to be *weakly preferred* over an outcome ω if $\omega' \geq \omega$. If in addition $\omega \not\geq \omega'$, then ω' is *strictly preferred* over ω, denoted $\omega' > \omega$. An agent is said to be *indifferent* between two outcomes if $\omega' \geq \omega$ and $\omega \geq \omega'$. In that case, we also say that these outcomes are *equally valued* and we write $\omega' \sim \omega$. An *indifference curve* or *iso-curve* is a set of outcomes that are equally valued by an agent. In a *total preference order*, one outcome is always (weakly) preferred over the other outcome for any outcome pair, which means there are no undefined preference relations. Finally, an outcome ω is *Pareto optimal* if there exists no outcome ω' that is preferred by an agent without making another agent worse off [23]. For two players A and B with respective preference orders $\geq_A$ and $\geq_B$, this means that there is no outcome ω' such that:

$$\left(\omega' >_A \omega \wedge \omega' \geq_B \omega \right) \vee \left(\omega' >_B \omega \wedge \omega' \geq_A \omega \right).$$

An outcome that is Pareto optimal is also said to be *Pareto efficient*. When an outcome is not Pareto efficient, there is potential, through re-negotiation, to reach a more preferred outcome for at least one of the agents without reducing the value for the other.

The outcome space can become quite large, which means it is usually not viable to explicitly state an agent's preference for every alternative. For this reason, there are more succinct preference representations for preferences [33, 34].

A well-known and compact way to represent preference orders is the formalism of conditional preference networks (CP-nets) [35–37]. CP-nets are graphical models, in which each node represents an negotiation issue and each edge denotes preferential dependency between issues. If there is an edge from issue i to issue j, the preferences for j depend on the specific value for issue i. To express conditional preferences, each issue is associated with a conditional preference table, which represents a total order of possible values for that issue, given its parents' values.

A preference profile may be specified as a list of ordering relations, but it is more common in the literature to express the agent's preferences by a *utility function*. A utility function assigns a utility value to every possible outcome, yielding a *cardinal* preference structure.

Cardinal preferences are 'richer' than ordinal preferences in the sense that ordinal preferences can only compare between different alternatives, while cardinal preferences allow for expressing the intensity of every preference [33]. Any cardinal preference induces an ordinal preference, as every utility function u defines an order $\omega' \geq \omega$ if and only if $u(\omega') \geq u(\omega)$.

Some learning techniques make additional assumptions about the structure of the utility function [38], the most common in negotiation being that the utility of a multi-issue outcome is calculated by means of a *linear additive function* that evaluates each issue separately [23, 38, 39]. Hence, the contribution of every issue to the utility is linear and does not depend on the values of other issues. The utility $u(\omega)$ of an outcome $\omega = \langle \omega_1, \ldots, \omega_n \rangle \in \Omega$ can be computed as a weighted sum from evaluation functions $e_i(\omega_i)$ as follows:

$$u(\omega) = \sum_{i=1}^{n} w_i \cdot e_i(\omega_i), \tag{2.1}$$

where the w_i are normalized weights (i.e. $\sum w_i = 1$). Linear additive utility functions make explicit that different issues can be of different importance to a negotiating agent and can be used to efficiently calculate the utility of a bid at the cost of expressive power, as they cannot represent interaction effects (or dependencies) between issues [36].

A common alternative is to make use of non-linear utility functions to capture more complex relations between offers at the cost of additional computational complexity. Non-linear negotiation is an emerging area within automated negotiation that considers multiple inter-dependent issues [40, 41]. Typically this leads to larger, richer outcome spaces in comparison to linear additive utility functions. A key factor in non-linear spaces is the ability of a negotiator to make a proper evaluation of a proposal, as the utility calculation of an offer might even prove NP-hard [42]. Examples of this type of work can be found in [43–46].

For non-linear utility functions in particular, a number of preference representations have been formulated to avoid listing the exponentially many alternatives with their utility assessment [33]. The utility of a deal can be expressed as the sum of the utility values of all the *constraints* (i.e., regions in the outcome space) that are satisfied [43, 47]. These constraints may in turn exhibit additional structure, such as being represented by hyper-graphs [48]. One can also decompose the utility function into *subclusters* of individual issues, such that the utility of an agreement is equal to the sum of the sub-utilities of different clusters [46]. This is a special case of a utility structure called *k-additivity*, in which the utility assigned to a deal can be represented as the sum of basic utilities of subsets with cardinality $\leq k$ [49]. For example, for $k = 2$, the utility $u(\omega_1, \omega_2, \omega_3)$ might be expressed as the utility value of the individual issues $u_1(\omega_1) + u_2(\omega_2) + u_3(\omega_3)$ (as in the linear additive case), plus their 2-way interaction effects $u_4(\omega_1, \omega_2) + u_5(\omega_1, \omega_3) + u_6(\omega_2, \omega_3)$. This is in turn closely related to the OR and XOR languages for bidding in auctions [50], in which the utility is specified for a specific set of clusters, together with rules on how to combine them into utility functions on the whole outcome space.

In our setting, both the domain and preferences stay fixed during a single negotiation encounter, but while the domain is common knowledge to the negotiating parties, the preferences of each player is private information. This means that the players do not have access to the utility function of the opponent. In more detail, even the opponent's orderings of the issues are unknown, and the agents are not provided with any prior distribution over the utility functions. However, the players can attempt to learn during the negotiation encounter.

The preference profile of an agent may also specify a *reservation value*. The reservation value is the minimal utility that the agent still deems an acceptable outcome. That is, the reservation value is equal to the utility of the best alternative to no agreement. A bid with a utility lower than the reservation value should not be offered or accepted by any rational agent. In a single-issue domain, the negotiation is often about the *price P* of a good [25, 27, 51, 52]. In that case, agent A and B usually take the roles of buyer B and seller S, and their reservation values are specified by their *reservation prices* RP_B and RP_S. RP_B denotes the highest price a buyer is willing to pay, while RP_S is the lowest price at which a seller is willing to sell.

The negotiator's nearness to a deadline is only one example of *time pressure* [53], which is defined as a negotiator's desire to end the negotiation quickly [54]. Another way to model time pressure is to supplement the negotiation scenario with a *discount factor*. Let d in [0, 1] be the discount factor and let t in [0, 1] be the current normalized time. We compute the discounted utility $u^d(\omega)$ from the undiscounted utility $u(\omega)$ as follows:

$$u^d(\omega) = u(\omega) \cdot d^t. \tag{2.2}$$

If $d = 1$, the utility is not affected by time, and such a scenario is considered to be undiscounted, while if d is very small, there is high pressure on the agents to reach an agreement. Note that discount factors are part of the scenario, are known to both agents and are always *symmetric* (i.e. d always has the same value for both agents).

The reasons for having deadlines and discount factors are both pragmatic and to make the negotiation more interesting from a theoretical perspective. Without a deadline or discount factor, the negotiators have no incentive to accept an offer, and so the negotiation might go on forever. Also, with unlimited time an agent may simply try a large number of proposals to learn the opponent's preferences. In addition, as opposed to having a fixed number of rounds, both the discount factor and deadline are measured in real time. This, in turn, introduces another factor of uncertainty since it is now unclear how many negotiation rounds there will be, and how much time an opponent requires to compute a counter offer. Also, this computational time will typically change depending on the size of the outcome space.

2.2.4 Outcome Spaces

A useful way to visualize the preferences of both players simultaneously is by means of an *outcome space plot* (Fig. 2.2). The axes of the outcome space plot represent the utilities of player A and B, and every possible outcome $\omega \in \Omega$ maps to a point $(u_A(\omega), u_B(\omega))$. The line that connects all of the Pareto optimal agreements is the Pareto frontier.

Note that the visualization of the outcome space together with the Pareto frontier is only possible from an *external* point of view. In particular, the agents themselves are not aware of the opponent utility of bids in the outcome space and do not know the location of the Pareto frontier.

From Fig. 2.2 we can immediately observe certain characteristics of the negotiation scenario. For example, the domain size, whether the bids are spread out over the domain, and the relative occurrence of Pareto optimal outcomes.

One important measure is the *bid distribution*, which is defined as the mean distance to the Pareto frontier. A scenario with a high bid distribution has a high

Fig. 2.2 A typical example of an outcome space between agents A and B

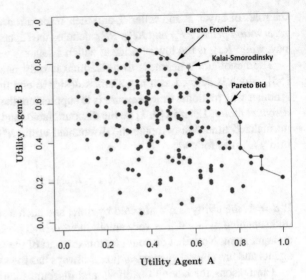

percentage of outcomes far from the Pareto frontier. This is defined formally as:

$$\text{distribution}(\Omega) = \sum_{\omega \in \Omega} \frac{\min_{p \in \Omega_P} d(\omega, p)}{|\Omega|}, \tag{2.3}$$

where $\Omega_P \subseteq \Omega$ is the set of Pareto efficient possible outcomes.

There are a number of special outcomes in the outcome space. Of course, the best result would be the outcome $\overline{\omega}$ at which both parties would receive their maximum utility. This would lead to complete satisfaction of both parties, but unfortunately, this is usually not a possible outcome.

There are also a number of definitions for what constitutes a *fair outcome* for both players [23]. The *Nash solution* is defined as the outcome that maximizes the product of the utilities of agents A and B:

$$\omega_{\text{Nash}} = \max_{\omega \in \Omega} u_A(\omega) \cdot u_B(\omega). \tag{2.4}$$

An alternative is the *Kalai-Smorodinsky solution*, which is defined as:

$$\omega_{\text{Kalai}} = \min_{\omega \in \Omega} \left(\frac{u_A(\overline{\omega})}{u_B(\overline{\omega})} - \frac{u_A(\omega)}{u_B(\omega)} \right). \tag{2.5}$$

The opposition of the negotiation scenario is determined by the minimum distance from the Kalai-Smorodinsky solution to the point $\overline{\omega}$.[1] Formally:

---

[1] There are various ways to define the opposition of a scenario (see [55]), but as in [56], we will employ a definition based on distance measures throughout the thesis. Another popular definition is: $\text{opposition}(\Omega) = \min_{\omega \in \Omega} d(\omega, \overline{\omega})$.

$$\text{opposition}(\Omega) = d(\omega_{\text{Kalai}}, \overline{\omega}) \qquad (2.6)$$

where Ω is the set of all possible outcomes, and $d(\omega_1, \omega_2)$ gives the Euclidean distance between two points ω_1, ω_2 in the outcome space, as defined in Eq. (2.7).

$$d(\omega_1, \omega_2) = \sqrt{(u_A(\omega_1) - u_A(\omega_2))^2 + (u_B(\omega_1) - u_B(\omega_2))^2}, \qquad (2.7)$$

When a gain for one party can be achieved only at a loss for the other party (i.e., when the preferences are *conflicting*), the negotiation scenario is said to be *competitive*, or to have *strong opposition*. Conversely, in a *cooperative* scenario (or: a scenario with *weak opposition*), both parties achieve either losses or gains simultaneously.

2.3 Negotiating Strategies

A negotiating agent employs a *negotiation strategy* to determine its action in a given negotiation state. Research on general agent negotiators has given rise to a broad variety of negotiation strategies that have already been established both in literature and in implementations, (e.g. [27, 57–61]). The strategies of the agents usually vary from equilibrium strategies in a game theoretical setting to more heuristic approaches. Here we focus in particular on self-interested, boundedly rational agents that are able to conduct bilateral negotiations with incomplete information (following the classification of [16]).

Examples of such general agent negotiators in the literature include, among others: Zeng and Sycara [52], who introduce a generic agent called *Bazaar*; Faratin et al. [58], who propose an agent that is able to make trade-offs in negotiations and is motivated by maximizing the joint utility of the outcome (that is, the agents are utility maximizers that seek Pareto-optimal agreements); Karp et al. [62], who take a game-theoretic view and propose a negotiation strategy based on game-trees; Jonker et al. [60], who propose a a concession oriented strategy called *ABMP*; and Lin et al. [63], who propose an agent negotiator called *QOAgent*.

The ANAC competition that we hosted brought forth an additional 60 advanced negotiation strategies (see Appendix B on ANAC and Appendices C–F for agent descriptions). Notable ANAC agent strategies include: *Agent K* [64, 65], which calculates its target utility based on the average and variance of previous bids and employs a sophisticated acceptance strategy; *IAMHaggler* [66–68], which uses Gaussian process regression technique to predict the opponent's behavior; *CUHK Agent* [69, 70], which adaptively adjusts its acceptance threshold based on domain and opponent analysis; *OMAC Agent* [71–74], which models the opponent using wavelet decomposition and cubic smoothing spline; *The Fawkes*, which combines the best bidding, learning, and accepting strategy components; and finally, *Meta-Agent* [75–77], which, for any given negotiation domain, dynamically selects the most successful ANAC agent to produce an offer.

In Chap. 3, we introduce a component-based architecture for negotiating agents, so we start by describing literature that investigates and evaluates such components. There are two categories of relevant work we highlight here: literature detailing the architecture of a negotiating agent's strategy (Sect. 2.3.1); and work that explores and combines a set of negotiation strategy components to find better strategies (Sect. 2.3.2).

Our component-based architecture consists of three basic components: a *bidding strategy*, which determines which concession should be made in a negotiation state; an *acceptance strategy*, which is used by an agent to determine whether an opponent's offer should be accepted; and optionally an *opponent model*, which can be used both by the *bidding strategy* and *acceptance strategy* to reach a better outcome by exploiting knowledge about the opponent. We provide some background on each of the components in Sects. 2.3.3–2.3.5.

2.3.1 Architecture of Negotiation Strategies

To our knowledge, there is little work in literature describing the generic components of a negotiation strategy architecture, at a similar level of detail as our BOA architecture, which is outlined in Chap. 3. For example, Bartolini et al. [78] and Dumas et al. [79] treat the negotiation strategy as a singular component.

Jonker et al. [60] present an agent architecture for multi attribute negotiation, where each component represents a specific process within the behavior of the agent, e.g.: attribute evaluation, bid utility determination, utility planning, and attribute planning. There are some similarities between the two architectures; for example, the utility planning and attribute planning component correspond to the bidding strategy component in our architecture. In contrast to our work however, Jonker et al. focus on tactics for finding a counter offer and do not discuss acceptance strategies. The fact that our architecture allows this, makes it possible to find better strategies to accept (see Chaps. 4 and 5).

Ashri et al. [80] introduce a general architecture for negotiation agents, discussing components that resemble our architecture; components such as a proposal evaluator and response generator resemble an acceptance condition and bidding strategy respectively. However, the negotiation strategy is described from a BDI-agent perspective (in terms of motivation and mental attitudes).

Hindriks et al. [81] introduce an architecture for negotiation agents in combination with a negotiation system architecture. Parts of the agent architecture correspond to our architecture, but they treat the acceptance strategy and bidding strategy as a singular component, and their focus is primarily on how the agent framework can be integrated into a larger system.

2.3.2 Negotiation Strategy Space Exploration

There are various ways to explore the automated negotiation strategy space by combining a set of negotiation strategies.

Faratin et al. [27] analyze the performance of pure negotiation tactics on single issue domains in a bilateral negotiation setting. The decision function of the pure tactic is then treated as a component around which the full strategy is built. While they discuss how tactics can be linearly combined, the performance of the combined tactics is not analyzed.

Some authors use genetic algorithms to automatically combine certain tactics or strategies. This approach is different to how we combine components using the BOA framework, however they do share certain traits, as they view a strategy consisting of different components and combine them in order to produce a better performing strategy. For example, Matos et al. [82] employ a set of baseline negotiation strategies that are time dependent, resource dependent, and behavior dependent [27], all with varying parameters. The negotiation strategies are encoded as chromosomes and combined linearly, after which they are used by a genetic algorithm to analyze the effectiveness of the strategies. The fitness of an agent is its score in a negotiation competition. This approach analyzes acceptance criteria that only specify a utility interval of acceptable values, and hence do not take time into account. The agents also do not employ explicit opponent modeling.

Eymann [83] also uses genetic algorithms with more complex negotiating strategies, evolving six parameters that influence the bidding strategy. The genetic algorithm uses the current negotiation strategy of the agent and the opponent strategy with the highest average income to create a new strategy, similar to other genetic algorithm approaches (see Beam and Segev [84] for a discussion of genetic algorithms in automated negotiation). The genetic algorithm approach mainly treats the negotiation strategy optimization as a search problem in which the parameters of a small set of strategies are tuned by a genetic algorithm. In Chap. 3, we analyze a more complex space of newly developed negotiation strategies, as our pool of surveyed negotiation strategies consists of strategies introduced in the ANAC competition, as well as the strategies discussed by Faratin et al. (see Sect. 2.3.3). Furthermore, our work combines different *components* instead of complete strategies or strategy parameters and also investigates the importance of particular components (see Chap. 10).

Ros and Sierra obtain promising results in [85] with a negotiation strategy that combines two components: a concession based strategy (either time-based or behavior-based [27]) that decreases a utility threshold to achieve an agreement, and a trade-off strategy [58] that searches for a satisfactory proposal. Our work in this thesis differs with Sierra et al. as we consider a much wider array of agents of which we are able to change the opponent model as well.

Finally, Ilany and Gal [75–77] take the approach of selecting the best strategy from a predefined set of agents, based on the characteristics of a domain. Through machine learning this agent is optimized to choose the best strategy for that particular domain. The difference with our work is that they combine whole strategies, whereas the BOA architecture combines the *components* of strategies. Our contribution is to define and implement an architecture that allows to easily vary all main components of a negotiating agent. Especially in Chap. 10, we study the effects of a much larger group of state of the art negotiation components than has been done before.

Another way to explore the space of negotiation strategies is to classify them according to their behavior. We do so in Chap. 8, in which we present a new classification method for negotiation strategies, based on their pattern of concession making. This chapter is inspired by ideas presented in [86] (of which parts originally appeared in unpublished work by Kersten in 2005). In [86], four dual negotiation orientations are distinguished, depending on the negotiator's own orientation and that of the negotiating partner. Both orientations can be either competitive or cooperative, leading to four different labels: *Competitor*, *Yielder*, *Exploiter*, and *Cooperator*. In Chap. 8, we re-use these labels to name the stance of a negotiator against different kinds of opponents. However in our work, the negotiators are assumed to have different responses to different observed behavior by the other party. Therefore, instead of the negotiator having one particular stance during the negotiation, the position of the negotiators can change in response to the competitiveness of the opponent. For example, a negotiator may be *both* an *Exploiter* (against a *Cooperator*), and a *Yielder* (against a *Competitor*). The negotiator would then be called an *Inverter*, as he takes on the reverse role of his opponent.

In [87], a classification scheme is given for electronic commerce negotiation, including characteristics of the negotiating agents. It is argued that agents can act in a self-interested way, or altruistically, or strike a balance in between. This choice is then seen as a component of the bidding strategy of the agent, which ultimately decides how and when to place offers, or when to withdraw, etc. Although the paper makes this distinction in bidding characteristics, it does not provide a definition or a way to quantify them.

Thomas [88] defines five conflict–handling modes that can be applied to negotiation: *competing*, *collaborating*, *compromising*, *avoiding*, and *accommodating*. Similar to our work in Chap. 8, the classification method uses two underlying dimensions. However, the underlying dimensions are different, namely: *assertiveness* (attempting to satisfy one's own concerns), and *cooperativeness* (attempting to satisfy other's concerns). This classification method is phrased in qualitative, intentional terms of the conflict-handler. Similarly, Zachariassen [89] distinguishes negotiation strategies into two strategy types: *distributive* and *integrative*. This description also focuses on the approach used by the negotiators. Our work has a different focus from both papers, centering around quantitative negotiation characteristics *in response* to agents having either high and low concession rates. Furthermore, we do not classify negotiation strategies in a binary way (either cooperative or non-cooperative), but we employ a continuous spectrum in our approach to classify the full space of negotiation strategies.

2.3.3 Bidding Strategies

The *bidding strategy*, also called *negotiation tactic* or *concession strategy*, is usually a complex strategy component. Two types of negotiation tactics are very common: *time-dependent tactics* and *behavior-dependent tactics*. Each tactic uses a *decision function*, which maps the negotiation state to a *target utility*. Next, the agent can search for a bid with a utility close to the target utility and offer this bid to the opponent (Fig. 2.3).

2.3.3.1 Time-Dependent Tactics

Functions which return an offer solely based on time are called *time-dependent tactics*. The standard time-dependent strategy calculates a target utility $u(t)$ at every turn, based on the current time t. Perhaps the most popular time-based decision function can be found in [20, 27], which, depending on the current normalized time $t \in [0, 1]$, makes a bid with utility closest to

$$u(t) = P_{min} + (P_{max} - P_{min}) \cdot (1 - F(t)), \qquad (2.8)$$

where

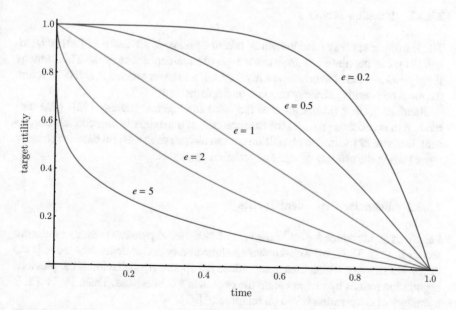

Fig. 2.3 Target utility through time of time-dependent tactics with concession factor $e \in \{0.2, 0.5, 1, 2, 5\}$

$$F(t) = k + (1 - k) \cdot t^{1/e}.$$

The constants $P_{\min}, P_{\max} \in [0, 1]$ control the range of the proposed offers, and $k \in [0, 1]$ determines the value of the first proposal. For $0 < e < 1$, the agent concedes only at the end of the negotiation and is called *Boulware*. If $e \geq 1$, the function concedes quickly to the reservation value, and the agent is then called a *Conceder*.

For $k = 0$ and $e = 1$, we obtain a very simple conceding tactic called *Conceder Linear*. It reduces Eq. (2.8) to

$$u(t) = P_{\min} + (P_{\max} - P_{\min}) \cdot (1 - t),$$

so that the agent linearly reduces its demanded utility from $P_{\max}$ to $P_{\min}$ as time passes.

In many of our experiments in later chapters, we set $k = 0$, and $P_{\max}, P_{\min}$ are respectively set to the maximum and minimum utility that can be obtained in the negotiation scenario. The specification of these strategies given in [20, 27] does not involve any opponent modeling; that is, given the target utility, a random bid is offered with a utility closest to it. Time-dependent tactics accept if and only if the opponent's bid is better than the target utility.

2.3.3.2 Baseline Tactics

The *Hardliner* strategy (also known as *take-it-or-leave-it*, *sit-and-wait* [90] or *Hard-ball* [91]) can be viewed as an extreme type of time-dependent tactic. This strategy simply makes a bid of maximum utility for itself and never concedes, and is therefore the most competitive strategy that can be implemented.

Random Walker (also known as the *Zero Intelligence* strategy [92]) generates random bids and thus provides the extreme case of a maximally unpredictable opponent. Because of its limited capabilities, it can also serve as a useful baseline strategy when testing the efficacy of other negotiation strategies.

2.3.3.3 Behavior-Dependent Tactics

Faratin et al. introduce a well-known set of *behavior-dependent tactics* or *imitative tactics* in [27]. The most well-known example of a behavior-dependent tactic is the *Tit for Tat* strategy, which tries to reproduce the opponent's behavior of the previous negotiation rounds by reciprocating the opponent's concessions. Thus, *Tit for Tat* is a strategy of cooperation based on reciprocity [93].

Tit for Tat has been applied and found successful in many other games, including the *Iterated Prisoner's Dilemma game* [94]. It is considered to be a very robust strategy, mainly because of the following three features:

1. It is never the first to defect (i.e., it plays nice as long as the the opponent plays nice as well);
2. It can be provoked into retaliation by a defection of the opponent;
3. However, it is forgiving after just one act of retaliation.

In total three tactics are defined: *Relative Tit for Tat*, *Random Absolute Tit for Tat*, and *Averaged Tit for Tat*. The *Relative Tit for Tat* agent mimics the opponent in a percentage-wise fashion by proportionally replicating the opponent's concession that was performed $\delta \geq 1$ steps ago. The decision function of *Relative Tit for Tat* is as follows:

$$x_{a \to b}^{t_{n+1}}[j] = \min \left(\max \left(\frac{x_{b \to a}^{t_n - 2\delta}[j]}{x_{b \to a}^{t_n - 2\delta + 2}[j]} x_{a \to b}^{t_{n-1}}[j], \min_j^a \right), \max_j^a \right) \qquad (2.9)$$

The formula specifies the value for each issue j for the next bid for the opponent $x_{a \to b}^{t_{n+1}}$ at time step t_{n+1}, and depends on the previous opponent offers $x_{b \to a}^{t_n - 2\delta}[j]$ and $x_{b \to a}^{t_n - 2\delta + 2}[j]$ in proportion to its own previous offer $x_{b \to a}^{t_{n-1}}[j]$. The *min* and *max* functions are used to ensure that the value of each issue stays within the acceptable range. The main weakness of the decision function is that a percentage concession by the opponent on a specific issue is in general unequal in utility compared to the same concession by the agent.

The standard *Tit for Tat* strategies from [27] do not employ any learning methods, but this work has been subsequently extended by the *Nice Tit for Tat agent* [95] and the *Nice Mirroring Strategy* [96]. These strategies achieve more effective results by combining a simple *Tit for Tat* response mechanism with learning techniques to propose offers closer to the Pareto frontier. These approaches can be viewed as simple examples of the ideas we explore in Sect. 3, where we study arbitrary combinations of concession strategies with learning methods.

2.3.4 Acceptance Strategies

All negotiation agent implementations have to deal with the question of when to accept. In many cases, the agent accepts a proposal when the value of the offered contract is higher than the offer it is ready to send out at that moment in time. This is a significant case, in which the bidding strategy effectively dictates the acceptance strategy. Examples include the time-dependent negotiation strategies defined in [85] (e.g. the *Boulware* and *Conceder* tactics). The same principle is used in the equilibrium strategies of [20] and for the *Trade-off agent* [58]. *Agent K* [64] employs a more sophisticated method to decide when to accept. Its *acceptance strategy* (or *acceptance mechanism*) is based on the mean and variance of all received offers. It then tries to determine the best offer it might receive in the future and sets its proposal target accordingly. We refer to the agent descriptions in C.1 and D.1 in the Appendix for more descriptions of acceptance strategies.

We treat acceptance mechanism design in more detail in Chap. 4, where we present a model for accepting offers and where we compare state-of-the-art acceptance conditions of a large set of negotiation strategies. Our negotiation model builds upon the model of [26], where one specific acceptance condition is studied. We take a more general approach in Chaps. 4 and 5, in which the agent utilizes a generic acceptance mechanism where the current time and the entire bidding history is considered.

We only consider the alternating offers protocol in this thesis, but there are multiple other accepting strategies available for other methods of reaching an agreement. In a multi-party setting, the problem of when to accept is more complex, as the outside options become dynamic; however, the presence of a mediator can reduce some of the complexity by taking over the role of finding acceptable agreements, for example through letting the agents vote on whether a proposed contract is acceptable [44]. It may then be sufficient for an agent to simply accept anything above its reservation value. In the same way, when richer protocols are employed (e.g., when communication is possible, for instance in persuasive, or argumentation-based negotiation [8, 97]), the acceptance dilemma may be easier to resolve, as agents have more knowledge about the acceptability of offers. Lastly, in traditional negotiation protocols such as alternating offers, once a contract is settled upon, it is binding. However, a more general approach is to allow decommitment, i.e. backing out of the negotiation after finding a superior option elsewhere, usually at the cost of a penalty [98]. This requires complex acceptance strategies for committing and decommitting to agreements in a concurrent way, which has recently opened up new research in this area [99–101].

2.3.4.1 Optimal Stopping

When we move from real-time negotiation to round-based negotiations, it becomes possible to adopt *optimal* acceptance strategies through backward inductive reasoning; the most well-known solution being that agreement is reached immediately in the first round [19]. In a real-time setting, it is generally unknown when the last offer has been made, and this makes it difficult to find optimal acceptance conditions for this setting.

In Chap. 5 we explore this idea, and we present the first work that deals with the optimal decision on the acceptance of an offer in a negotiation setting of incomplete information. In many settings of *complete* information ([19] is a typical example) the deal is usually formed right away and as such, sequential decisions whether to accept do not come into play. In [52], a sequential decision making framework is also employed, using similar arguments for using it as we do. Furthermore, they also choose actions that maximize the expected payoff using a recursive formula; however, their approach uses Bayesian learning techniques and does not provide solutions specifically aimed at acceptance strategies. The work by Fatima et al. [25] also treats optimal strategies in an incomplete information setting, but it primarily focuses on bidding strategies in the context of unknown deadlines and reservation values, and does not deal with acceptance strategies. Research that comes closest to our work on optimal acceptance strategies is presented in [31], where optimal stopping is employed to decide when a party should reach an agreement in the context of conflict resolution. In contrast to our work, the scope of the paper is limited to simple bargaining games, and deals with one-sided incomplete information only.

We come back to optimal stopping and sequential decision making in Chap. 9 when we formulate optimal concession curves. To the best of our knowledge, that is the first work that makes usage of the optimal stopping rule to *generate* offers in an incomplete information setting and compares it to other concession techniques, where other previous work makes use of optimal stopping theory to formulate acceptance strategies in different settings [102–104]; for instance deciding to accept sequential job offers while trying to maximize the sum of the payments of all accepted jobs [104]. The major difference with optimal acceptance policies and our work in Chap. 9 is that we use the optimal stopping rule for *concessions*, instead of focusing on the optimal time to accept. Our work in Chap. 9 is defined more as the *complimentary* version of our approach in Chap. 5, in the sense that our formulation of optimal bidding rules happen to resemble optimal acceptance rules. Another key point is that we do not assume that the players' strategies are fixed, which allows us to formulate optimal bidding strategies against certain types of accepting strategies.

2.3.5 Opponent Models

An opponent model is an abstracted description of a player (and/or of a player's behavior) during the game [105]. There are many different types of opponent models; for instance, a model can describe the opponent's preferences, strategy, weaknesses, knowledge, and so on. We present here a short background on learning techniques and evaluation techniques in negotiation for our setting; for a more detailed exposition we refer to our survey on this topic [106].

In negotiation, opponent modeling often revolves around three questions:

- **Preference estimation**. What does the opponent want?
- **Strategy prediction**. What will the opponent do, and when?
- **Opponent classification**. What kind of player is the opponent, and how should we act accordingly?

The above questions are often highly related. For example, some form of preference estimation is needed in order to adequately interpret the opponent's actions. Then, knowing how the opponent acted according to its own utility, we can deduce its strategy, which in turn can help predict what the agent will do in the future. We will mainly focus on preference modeling in this thesis, although our architecture can accommodate for the other types of opponent models as well (see Chaps. 6 and 7).

Constructing an opponent model may alternatively be viewed as a *classification problem* where the *type* of the opponent needs to be determined from a range of possibilities [107]; one example being the work by Lin et al. [63]. Here the *type* of an opponent refers to all opponent attributes that may be modeled to gain an advantage in the game.

Opponent modeling can be performed online or offline, depending on the availability of historical data. Offline models are created before the negotiation starts, using previously obtained data from earlier negotiations. Online models are constructed from knowledge that is collected during a single negotiation session, which is the focus of this thesis. A major challenge in online opponent modeling is that the model needs to be constructed from a limited amount of exchanged bids, and real-time deadlines may pose the additional challenge of having to construct the model as fast as possible. Even though there are large differences between the models, a common set of high level motivations behind their construction can be identified. There are the following motivations for why opponent models are used in automated negotiation:

1. **Augment behavior-based tactics** [95, 96, 108, 109] An opponent model can assist in improving the performance of behavior-based tactics, as the opponent's concessions can then be estimated and reciprocated more accurately. Based on move classification, behavior-based strategies such as the *Tit for Tat* strategy (see Sect. 2.3.3.3) can be applied. In addition, in a negotiation where the opponent's preferences are private, an agent's concession might accidentally result in a decrease in utility for the opponent as well. Such an offer is called *unfortunate* [108], and can be avoided by better estimating the opponent's preferences.

2. **Avoid non-agreement** [28, 110–120] In most negotiations, reaching an agreement is preferred over not reaching a deal. The opponent's previous moves can be analyzed to estimate the minimal concessions required to ensure acceptance.

3. **Find a counter-strategy** [68, 69, 72, 73, 110, 111, 121–133] The opponent can be exploited in multiple ways with the assistance of an opponent model. One way is to estimate the opponent's reservation value in an attempt to obtain the minimal negotiation outcome the opponent will settle for. Alternatively, an estimate of the opponent's deadline can be used to elicit concessions from the opponent by stalling the negotiation, provided of course that the agent has a later deadline. Theoretical results are available that specify which counter-strategy to use depending on the information known about the opponent [25, 82, 134].

4. **Maximize social welfare** [46, 51, 52, 63, 112–115, 118–120, 135–141] In a cooperative environment, agents aim for a fair result. An agent can use an estimate of the opponent's preference profile to maximize the chances of a good outcome for both.

5. **Propose Pareto optimal bids** [28, 46, 67, 95, 96, 109, 118, 135, 137, 140, 142–152] Pareto optimality of an offer ensures the offer cannot be improved for both players at the same time. When an agent considers multiple similarly preferred offers to send out to the opponent, offering a Pareto optimal bid can lead to an earlier and mutually beneficial agreement.

6. **Reduce negotiation costs** [46, 51, 52, 109, 110, 113, 115, 119, 120, 135–137, 139–141, 148, 153–155] In general it costs time and resources to negotiate, and using an estimate of the opponent's preference profile or negotiation strategy

can aid in reducing these costs. An agent may even decide that the estimated negotiation costs are too high to warrant a potential agreement, and prematurely end the negotiation.

We found that existing work on opponent models can fulfill any of the goals above by learning a combination of *six* opponent attributes, which we have listed in Table 2.1. The notion of an opponent model as a component of a negotiation strategy has been discussed by many of these authors. However, to our knowledge, there is limited work in which the performance of different types of opponent models is compared as we do in Chaps. 6 and 7. One example is the work by Papaioannou et al. [116], who evaluate a set of opponent strategy prediction techniques in terms of resulting performance gain.

2.4 Evaluation Methodologies

We now introduce the methodologies we use in subsequent chapters to evaluate negotiation strategies. The first evaluation method is analytical software to analyze the performance and dynamics of agents, and the outcome of the negotiation (Sect. 2.4.1). The second is a method to benchmark and objectively evaluate negotiation agents in a competitive setting (Sect. 2.4.2). Together, they provide an environment to apply a range of *performance measures* (Sect. 2.4.3) to measure the performance of a negotiation strategy. Lastly, we discuss measures for learning methods, including *accuracy measures* (Sect. 2.4.4).

2.4.1 Environments for Evaluating Negotiating Agents

As we have built a generic environment for designing and evaluating agent negotiators called GENIUS [165] (see Appendix A), we briefly review related work that is explicitly aimed at the evaluation of various agent negotiators. Most of the work reported herein concerns the evaluation of various *strategies* for negotiation used by such agents. Although some results were obtained by game-theoretic analysis (e.g. [7, 15]), most results were obtained by means of *simulation* (e.g. [166–168]). Devaux and Paraschiv [166] present work that compares agents negotiating in internet agent-based markets. In particular, they compare a strategy of their own agent with behavioral based strategies taken from the literature [27]. The simulations are performed in an abstract domain where agents need to negotiate the price of a product. Similarly, Henderson et al. [168] present results of the performance of various negotiation strategies in a simulated car hire scenario. Finally, Matos et al. [82] conducted experiments to determine the most successful strategies using an evolutionary approach in an abstract domain called the *service-oriented domain*. Even though several of the approaches use an abstract domain with a range of parameters

Table 2.1 An overview of learning techniques and methods that help to learn six different opponent attributes

Opponent attributes	Procedure	Learning techniques
Reservation value	Bidding strategy estimation	Bayesian learning [51, 52, 110, 119, 120, 130, 133, 141]
		Non-linear regression [111, 121, 125, 133]
Deadline	Bidding strategy estimation	Bayesian learning [110, 119, 133]
		Non-linear regression [110, 119, 125, 133]
Issue preference order	Measuring similarity between offers	Bayesian learning [28]
		Kernel density estimation [136, 156]
		Heuristics [143, 147]
	Knowledge of bidding strategy	Simplified genetic algorithm [112]
Outcome preference order	Classification	Bayesian learning [63, 67, 95, 109, 138, 146, 150, 157, 158]
	Data mining aggregate preferences	Random variable estimation [113, 140]
		Graph theory [46, 148]
		Bayesian network [149]
	Logical reasoning and heuristics	Heuristics [69, 144, 145, 151–155, 159]
Bidding strategy	Regression analysis	Non-linear regression [111, 116, 121, 123, 125, 133, 160]
		Polynomial interpolation [116]
		Genetic algorithms [117]
		Bayesian networks [127]
	Time series forecasting	Derivatives [122, 139]
		Signal processing [68, 72, 73, 129, 132, 161]
		Neural networks [114, 116–118, 126, 162–164]
		Markov chains [128]
Acceptance strategy	Interpolation of acceptance probability	Polynomial interpolation [131]
		Kernel density estimation [115]
		Bayesian learning [137]
		Neural networks [124]

that may be varied, we argue that the focus on a single domain in most simulations is restrictive. A similar argument to this end has been put forward in [56]. The analysis of agent negotiators in multiple domains may significantly improve the performance of such agents.

Manistersky et al. [169] discuss how people who design agent negotiators change their design over time. They study how students changed their design of a trading agent that negotiates in an open environment. After initial design of their agents, human designers obtained additional information about the performance of their agents by receiving logs of negotiations between their agents and agents designed by others. These logs provided the means to analyze the negotiation behavior, and an opportunity to improve the performance of the agents. The GENIUS environment discussed in Appendix A provides a tool that supports such analysis, subsequent improvement of the design, and structures the enhancement process.

Part of GENIUS' functionality has been described in [170, 171], and our work [165] outlined in Appendix A is a natural extension of this research. Since then, we have extended GENIUS with all ANAC resources and new functionality described in Appendix B (e.g., negotiation strategies, protocols, scenarios, discount factors, reservation values), the BOA architecture and agent components from Chap. 3, the acceptance strategies from Chaps. 4 and 5, and the performance and accuracy measures described in Chaps. 6 and 7.

With regard to systems that facilitate the actual design of agents or agent strategies in negotiations, few systems are close to GENIUS. Most of the systems that may be related to its main focus are negotiation support systems (e.g., the Interactive Computer-Assisted Negotiation Support system (ICANS) presented in [172], the InterNeg Support Program for Intercultural REsearch (INSPIRE)), however, GENIUS advances the state-of-the-art by also providing evaluation mechanisms that allow a quick and simple evaluation of strategies and the facilitation of automated negotiator's design. INSPIRE, by Kersten and Noronha [173], is a Web-based negotiation support system with the primary goal of facilitating negotiation research in an international setting. The system enables negotiation between two people, collects data about negotiations and has some basic functionality for the analysis of the agreements, such as calculation of the utility of an agreement and exchanged offers. However, unlike GENIUS, it does not allow integration of an automated negotiating agent and thus does not include repositories of agents as we propose. Perhaps Neg-o-Net [174] is more similar to GENIUS than all the other support systems. The Neg-o-Net model is a generic agent-based computational simulation model for capturing multi-agent negotiations concerning resource and environmental management decisions. The Neg-o-Net model includes both a negotiation algorithm and some agent models. An agent's preferences are modeled using digraphs (scripts). Nodes represent states of the agent that can be achieved by performing actions (arcs). Each state is evaluated using utility functions. The user can modify the agent's script to model his/her preferences w.r.t. states and actions. While Neg-o-Net is similar to GENIUS, there are at least two important differences. First, they currently do not support the incorporation of human negotiators, but only automated ones. Second, they do not provide any evaluation mechanism of the strategies as GENIUS provides.

A recent development worth noting is the *Negowiki* project [17, 175], which aims to unify current approaches in negotiation research by creating a collection of standardized negotiation scenarios. *Negowiki* is an online framework where researchers can share negotiation scenarios and results. As in GENIUS, analysis of the results is provided, so that researchers can compute a set of metrics over the results of the negotiation (e.g. Pareto optimality, fairness; we elaborate more on this in Sect. 2.4.3). All scenarios offered by *Negowiki* are also available for download in GENIUS format.

2.4.2 Negotiating Agent Competitions

A *competition* can act as a useful and open benchmarking tool to evaluate and compare negotiation agents, as evidenced by successful competitions to advance the state-of-the-art in artificial intelligence such as the *Computer Poker Competition* [176], the Iterated Prisoner's Dilemma game [94] and the *Trading Agent Competition* [177]. Following in their footsteps, we organized four annual instances of the International Automated Negotiating Agents Competition (ANAC).

We elaborate on the goals and results of the competition in Appendix B. Here, we provide a short description of related competitions and outline the differences with ANAC.

2.4.2.1 The Trading Agent Competition

Four games of the *Trading Agent Competition* (TAC) relate to automated negotiating agents [177–181], and some elements of TAC have similar challenges as posed by ANAC:

TAC SCM *TAC Supply Chain Management* was designed to simulate a dynamic supply chain environment. Agents have to compete to secure customer orders and components required for production. In order to do so, the agents have to plan and coordinate their activities across the supply chain. Participants face the complexities of supply chains, which admits a variety of bidding and negotiation strategies.

TAC Ad Auctions In the *TAC Ad Auctions*, game entrants design and implement bidding strategies for advertisers in a simulated sponsoring environment. The agents have to bid against each other to get an ad placement that is related to certain keyword combinations in a web search tool. The advertiser strategies have to decide which keywords to bid on, and what prices to offer. Therefore, the strategies have to optimize their data analysis and bidding tactics to maximize their profit.

TAC Market Design *TAC Market Design* or *The CAT Competition* is a reverse of the normal TAC game: as an entrant you define the rules for matching buyers and sellers, while the trading agents are created by the organizers of the competition. Entrants have to compete against each other to build a robust market mechanism that attracts buyers and sellers.

Power TAC Having started in 2011, *Power TAC* is a fairly recent addition to the TAC games. It is built around a competitive market simulation platform with the goal to direct policy making and to develop and validate intelligent agent technology for trading. It models a electrical energy market, where competing business entities offer energy services to customers.

The challenges posed by TAC are similar as in ANAC, especially the games of *TAC Ad Auctions* and *Power TAC*. The games of TAC can get very complex and the domains of the games are specifically chosen to model a certain scenario of a trading agent problem. Contrastingly, the entrants of ANAC have to consider very generic negotiation domains when they design their agents. On the one hand, this makes ANAC very accessible, as there are no domain-dependent details the participants have to know about. On the other hand, it is very difficult to develop an agent that negotiates well under such a wide variety of circumstances, especially with the unique challenges ANAC poses, which include one-shot bilateral negotiations with a real timeline, combined with incomplete information of the opponent's preferences.

2.4.2.2 The Agent Reputation Trust Competition

The *Agent Reputation Trust Competition* (ART) [182, 183] is also a negotiating agent competition with a testbed that allows the comparison of different strategies. The ART competition simulates a business environment for software agents that use the reputation concept to buy advices about paintings. Each agent in the game is a service provider responsible for selling its opinions when requested. The agent can exchange information with other agents to improve the quality of their appraisals. The challenge is to perceive when an agent can be trusted and to establish a trustworthy reputation. Compared to ANAC, the focus of ART is more on trust: the goal is to perceive which agents can be trusted in a negotiation process and what reputation should be attributed to each agent.

2.4.3 *Evaluating Performance of Negotiation Strategies*

The ultimate aim of a negotiation strategy is to increase overall performance of the negotiation, which is why *performance measures* are used to evaluate a negotiator's success. Performance measures evaluate the quality of the outcome, usually measured in utility gain, or distance of the agreement to the Pareto frontier. With this method, the success of an opponent model is expressed in terms of the negotiation *result* (as opposed to the whole negotiation *process*; for this we refer to Sect. 2.4.4) The paragraphs below provide an overview of the performance measures in related work.

Average utility. Average utility is by far the most popular performance measure and is used by many authors (e.g., [28, 63, 69, 72, 73, 96, 108–111, 115, 117, 119, 120, 122–126, 128, 129, 131–133, 135, 136, 138, 140–142, 146, 147, 150, 151,

156]). A common application is to consider the average utility of an agent with and without opponent model against a group of opponents on several domains (see for example [28, 69, 73]). Note that the average utility of an agent directly depends on the negotiation setting (as we will see in following chapters), which therefore should be chosen with care.

Distance to a fair outcome. Other authors are concerned with achieving a fair outcome [96, 108, 135, 139], which is especially important if there will be future negotiations between the parties.

Distance to a fair outcome is then calculated as the average Euclidean distance to a fair solution (as defined in Sect. 2.2.4), such as the distance to Nash solution [96, 108, 135, 139] or distance to Kalai-Smorodinsky [96, 108, 135]. As with the average utility measure, the negotiation setting strongly influences the result [184, 185].

Distance to Pareto frontier. An opponent model of the opponent's preferences aids in identifying Pareto optimal bids. For this type of model—assuming it is applied by a bidding strategy that takes the opponent's utility into account—the distance to the nearest Pareto optimal bid directly correlates with the model's quality (see for example [28, 109, 118, 135, 137]). Minimizing this distance to the Pareto-optimal frontier improves fairness and the probability of acceptance.

Joint utility. An alternative method to measure the fairness of an outcome is to calculate the joint utility [51, 52, 63, 112–115, 118–120, 136–139, 141]. The majority of the authors simply use the sum of the utility of the final outcome for the agents (see for example [63, 138]). An alternative used by several authors [51, 52, 114, 141] is to consider the normalized joint utility:

$$u_{\text{joint}} = \frac{(P - RP_S)(RP_B - P)}{(RP_B - RP_S)^2}. \tag{2.10}$$

In this equation, P is the agreed upon price, and RP_B and RP_S are the reservation prices of the buyer and seller respectively. Note that this definition is only applicable to single-issue negotiations. An alternative measure for multi-issue negotiations used by Jazayeriy et al. [112] is the geometric mean:

$$u_{\text{joint}} = \sqrt{u_A \cdot u_B}, \tag{2.11}$$

where u_A and u_B are the utilities achieved by the agents. An attractive property of this metric is that when the utilities are highly unbalanced, this formula better reflects unfairness than by simply calculating the sum of the utilities.

Percentage of agreements. An opponent model may lead to better bids being offered to the opponent, possibly avoiding non-agreement. In situations where an agreement is always better than no agreement, the percentage of agreements is a direct measure of success (see for example [28, 46, 110–121, 127, 133, 140, 144, 148, 155]). An important disadvantage is that the acceptance ratio does not capture the quality of the agreement. Agrawal and Chari, Buffett et al., and Mudgal and Vassileva use a related measure in which they calculate how often one agent outperforms the other with regard to the final outcome [121, 127, 144]. A disadvantage of this method is that an agent might outperform other agents, but still reach a bad outcome. An alternative metric is applied by Robu and Poutré [46, 148], which calculates how often an outcome is reached that maximizes social welfare.

Time of agreement. Various authors measure the duration of the negotiation or the communication load (e.g. [46, 51, 52, 109, 110, 113–115, 119, 120, 135–137, 139–141, 148, 153–155]), because in practical settings there is often a non-negligible cost associated with both. Opponent models can lead to earlier agreements, and thereby reduce costs. An important disadvantage of this metric is that while an opponent model may lead to an earlier agreement, the quality of the outcome for the agent might be lower.

Trajectory analysis. The quality of bidding strategies can be measured by analyzing the percentage and relative frequency of certain types of moves [96, 108, 109]. For example, unfortunate moves are offers that decrease the utility for both agents at the same time. Theoretically, a perfect opponent model of the opponent's preferences would allow an agent to prevent any such unfortunate moves. A disadvantage of this method is that it highly depends on the concession strategy that is used in combination with the opponent model.

2.4.4 Evaluating Learning Methods

The performance measures discussed in Sect. 2.3.5 are benchmarks for an entire negotiation strategy, but they are also often used to test the efficacy of one specific component, the most prevalent being the learning component. The simplest approach is to compare a novel learning technique with a set of baseline strategies. In [146] for example, the performance of the opponent model is estimated by embedding it in a strategy and comparing the average utility against two baseline strategies. The modeling technique discussed by [63] introduces a model for a similar protocol, but in this case the baseline is set by humans. Zeng and Sycara [52] measure performance in terms of social welfare, but focus on single-issue negotiations in which they compare the performance of three settings: both learn, neither learn, and only the buyer learns. Finally, [158] evaluates the accuracy of a model against simple baseline strategies in terms of the likelihood that the correct class is estimated to which the opponent's preference profile belongs.

The performance of an opponent model can also be tested against other models or against a theoretical lower or upper bound, as we do in Chap. 6. For example, Coehoorn and Jennings [136] evaluate the performance of their opponent model using a standard bidding strategy that can be used both with and without a model. The performance of the strategy is evaluated in three settings: without knowledge, with perfect knowledge, and when using an offline opponent model. This work is similar to our work in Chap. 6, however, it differs in the fact that we focus on *online* opponent modeling, and our setting is especially challenging as it involves the time/exploration trade-off. Another example is the work by [113], which introduces two opponent models for e-recommendation in a multi-object negotiation. Finally, [56] defines two accuracy measures and uses these measures to analyze the accuracy of two opponent models. The main differences are that in Chap. 6, we focus on the more general type of multi-issue negotiations, we focus on a larger set of performance measures, and pay more attention to the factors that influence the performance of the model. Furthermore, as far as we know, our work is the first to compare and analyze such a large set of state-of-the art models of the opponent's preference profile.

2.4.4.1 Accuracy Measures

As performance measures are only indirect measures of the negotiating agent's quality, other measures, such as *accuracy measures* can also be included for the purpose of benchmarking learning techniques. Accuracy measures are direct measures of learning quality, as they quantify the difference between the *estimate* and the *estimated*; i.e., they determine the quality of a model by quantifying how well the opponent model *represents* the real preferences of the opponent. An example is the correlation between the estimated and the real outcome space, or the percentage of correctly inferred Pareto optimal outcomes. We describe here the accuracy measures for preference modeling methods, as we come back to them in Chap. 7, where we compare the accuracy of various preference modeling techniques, using established accuracy measures.

For example, Carbonneau et al. [162] calculate the Pearson correlation between the real and estimated utility of the opponent's next bid. Hindriks and Tykhonov [56] extend this approach by measuring the Pearson correlation of the whole outcome space and discuss analogous definitions for the ranking distance. Our method in Chap. 7 incorporates both measures. An alternative approach is to measure the distance between elements of two preference profiles. For example Jazayeriy et al. [112] introduce such measures for the learning error of issue weights. We have incorporated these measures in our method, and we also apply the same measures to quantify the similarity between two full bid spaces.

Finally, there exist accuracy measures tailored to specific learning methods. Buffett and Spencer [157] for example, define a metric for opponent models that use Bayesian learning. The measure is defined as the average likelihood that the correct hypothesis is chosen from the set of candidate hypotheses. Since we employ models in Chaps. 6 and 7 that are based on a wide range of learning techniques, we do not incorporate measures specific to a particular learning method.

We also quantify the *relationship between accuracy and performance* in Chap. 7. In related work by Coehoorn and Jennings [136], a model is introduced that estimates the opponent's issue weights and the influence of small prediction errors on performance is investigated. The method in this thesis takes this a step further, as we analyze the relation between an exhaustive set of accuracy measures—including accuracy of the issue weights—and performance.

Similarity between issue weights We can measure the accuracy of models that estimate the issue weights of the opponent's preference profile in several ways [56, 112, 142]. All of them use a distance metric between the issue weights $w = (w_1, \ldots, w_n)$ of the real opponent preferences u_{op} and the issue weights $w' = (w'_1, \ldots, w'_n)$ of the estimated preferences u'_{op}. One way to do so is to measure the distance between the issue weight vectors [112]:

$$d_{\text{Euclidean}}(w, w') = \sqrt{\sum_{i=1}^{n}(w_i - w'_i)^2}. \tag{2.12}$$

Of course, this measure can be used for scalars as well. When modeling the opponent's deadline (or reservation value) $x \in \mathbb{R}$ with an estimate x', Eq. (2.12) simplifies to

$$d_{\text{Euclidean}}(x, x') = |x - x'|. \tag{2.13}$$

Another way is to check whether the issue weights are *ranked* correctly [56] by evaluating all possible pairs of issues $i_1, \ldots, i_n$:

$$d_{\text{rank}}(w, w') = \frac{1}{n^2}\sum_{j=1}^{n}\sum_{k=1}^{n} c(i_k, i_j), \tag{2.14}$$

where $c(i_k, i_j)$ is the conflict indicator function, which is equal to one when the ranking of the weights of issues i_k and i_j differs between the two profiles, and zero otherwise. An alternative is to measure the *correlation* between the vectors [56]:

$$d_{\text{Pearson}}(w, w') = \frac{\sum_{i=1}^{n}(w_i - \overline{w})(w'_i - \overline{w'})}{\sqrt{\sum_{i=1}^{n}(w_i - \overline{w})^2 \sum_{i=1}^{n}(w'_i - \overline{w'})^2}}. \tag{2.15}$$

Note that this expression may be undefined, for example when all weights are equal.

Similarity between preference profiles When opponent models estimate the opponent's preferences fully (e.g. [56, 142, 148, 157–159]), the quality of these models depends on the similarity between the real u_{op} and estimated opponent's preference profile u'_{op} for all bids in the outcome space Ω. One approach is to calculate the *average distance* between all outcomes in Ω [142]:

$$d_{\text{abs}}(u_{op}, u'_{op}) = \frac{1}{|\Omega|}\sum_{\omega \in \Omega} |\omega - \omega'|. \tag{2.16}$$

However, in practice, the correct ranking of the bids can be sufficient already. An alternative is therefore to use the *ranking distance of bids* measure that compares all preference orderings in a pairwise fashion [56]:

$$d_{\text{rank}}(u_{op}, u'_{op}) = \frac{1}{|\Omega|^2} \sum_{\omega \in \Omega, \omega' \in \Omega} c_{\prec u, \prec u'}(\omega, \omega'), \qquad (2.17)$$

where $c_{\prec u, \prec u'}$ is the conflict indicator function, which is equal to one when the ranking of the outcomes ω and ω' differs between the two profiles, and zero otherwise. Identically, Buffett et al. count the amount of correctly estimated preference relations [157, 158]. A disadvantage of these approaches is their scalability because all possible outcome pairs need to be compared. This problem can be overcome by using a Monte Carlo simulation; however, a more efficient solution can be to use the *Pearson correlation of bids* [56], which is defined as follows:

$$d_{\text{Pearson}}(u_{op}, u'_{op}) = \frac{\displaystyle\sum_{\omega \in \Omega}(u_{op}(\omega) - \overline{u_{op}})(u'_{op}(\omega) - \overline{u'_{op}})}{\sqrt{\displaystyle\sum_{\omega \in \Omega}(u_{op}(\omega) - \overline{u_{op}})^2 \sum_{\omega \in \Omega}(u'_{op}(\omega) - \overline{u'_{op}})^2}} \qquad (2.18)$$

A downside of this measure, although unlikely to occur in practice, is that it is not defined for all inputs, for example when all bids are estimated to have the same utility.

This chapter is based on the following publications: [95, 106, 165, 184–188]

Tim Baarslag, Koen V. Hindriks, and Catholijn M. Jonker. A tit for tat negotiation strategy for real-time bilateral negotiations. In Takayuki Ito, Minjie Zhang, Valentin Robu, and Tokuro Matsuo, editors, *Complex Automated Negotiations: Theories, Models, and Software Competitions*, volume 435 of *Studies in Computational Intelligence*, pages 229–233. Springer Berlin Heidelberg, 2013

Tim Baarslag, Mark J.C. Hendrikx, Koen V. Hindriks, and Catholijn M. Jonker. Learning about the opponent in automated bilateral negotiation: a comprehensive survey of opponent modeling techniques. *Autonomous Agents and Multi-Agent Systems*, pages 1–50, 2015

Raz Lin, Sarit Kraus, Tim Baarslag, Dmytro Tykhonov, Koen V. Hindriks, and Catholijn M. Jonker. Genius: An integrated environment for supporting the design of generic automated negotiators. *Computational Intelligence*, 30(1):48–70, 2014

Tim Baarslag, Katsuhide Fujita, Enrico H. Gerding, Koen V. Hindriks, Takayuki Ito, Nicholas R. Jennings, Catholijn M. Jonker, Sarit Kraus, Raz Lin, Valentin Robu, and Colin R. Williams. Evaluating practical negotiating agents: Results and analysis of the 2011 international competition. *Artificial Intelligence*, 198:73–103, May 2013

Tim Baarslag, Koen V. Hindriks, Catholijn M. Jonker, Sarit Kraus, and Raz Lin. The first automated negotiating agents competition (ANAC 2010). In Takayuki Ito, Minjie Zhang, Valentin Robu, Shaheen Fatima, and Tokuro Matsuo, editors, *New Trends in Agent-based Complex Automated Negotiations*, volume 383 of *Studies in Computational Intelligence*, pages 113–135, Berlin, Heidelberg, 2012. Springer-Verlag

Tim Baarslag, Koen V. Hindriks, Mark J.C. Hendrikx, Alex S.Y. Dirkzwager, and Catholijn M. Jonker. Decoupling negotiating agents to explore the space of negotiation strategies. In Ivan Marsa-

Maestre, Miguel A. Lopez-Carmona, Takayuki Ito, Minjie Zhang, Quan Bai, and Katsuhide Fujita, editors, *Novel Insights in Agent-based Complex Automated Negotiation*, volume 535 of *Studies in Computational Intelligence*, pages 61–83. Springer, Japan, 2014

Tim Baarslag, Koen V. Hindriks, and Catholijn M. Jonker. Towards a quantitative concession-based classification method of negotiation strategies. In David Kinny, Jane Yung-jen Hsu, Guido Governatori, and Aditya K. Ghose, editors, *Agents in Principle, Agents in Practice*, volume 7047 of *Lecture Notes in Computer Science*, pages 143–158, Berlin, Heidelberg, 2011. Springer Berlin Heidelberg

Tim Baarslag and Koen V. Hindriks. Accepting optimally in automated negotiation with incomplete information. In *Proceedings of the 2013 International Conference on Autonomous Agents and Multi-agent Systems*, AAMAS '13, pages 715–722, Richland, SC, 2013. International Foundation for Autonomous Agents and Multiagent Systems

References

1. Kraus S, Wilkenfeld J, Harris MA, Blake E (1992) The hostage crisis simulation. Simul Gaming 23(4):398–416
2. Sandholm T, Lesser VR (1995) Issues in automated negotiation and electronic commerce: Extending the contract net framework. In: Proceedings of the first international conference on multi-agent systems (ICMAS). San Francisco, pp 328–335
3. Shehory O, Kraus S (1998) Methods for task allocation via agent coalition formation. Artif Intell 101(1–2):165–200
4. Smith RG (1980) The contract net protocol: High-level communication and control in adistributed problem solver. IEEE Trans Comput 29(12):1104–1113
5. Deutsch M, Coleman PT, Marcus EC (2000) The handbook of conflict resolution: theory and practice, 1st edn. Jossey-Bass, San Francisco
6. Zlotkin G, Rosenschein JS (1991) Cooperation and conflict resolution via negotiation among autonomous agents in noncooperative domains. IEEE Trans Syst Man Cybern 21(6):1317–1324
7. Kraus S (2001) Strategic negotiation in multiagent environments. MIT Press, Cambridge
8. Sycara KP (1993) Machine learning for intelligent support of conflict resolution. Decis Support Syst 10(2):121–136
9. Jennings NR, Faratin P, Lomuscio AR, Parsons S, Wooldridge MJ, Sierra C (2001) Automated negotiation: Prospects, methods and challenges. Group Decis Negot 10(2):199–215
10. Sycara KP (1985) Arguments of persuasion in labour mediation. Proceedings of the 9th international joint conference on artificial intelligence, vol 1. Morgan Kaufmann Publishers Inc, San Francisco, pp 294–296
11. Sycara KP (1988) Resolving goal conflicts via negotiation. In: Proceedings of the 7th national conference on artificial intelligence. St. Paul, MN, pp 245–250, 21–26 August 1988
12. Robinson WN (1990) Negotiation behavior during requirement specification. In: Proceedings of the 12th international conference on software engineering, pp 268–276
13. Rosenschein JS (1996) Rational interaction: cooperation among intelligent agents. Ph.D. thesis, Stanford University, Stanford, 1986
14. Klein M, Lu SC-Y (1989) Conflict resolution in cooperative design. Artif Intell Eng 4(4):168–180
15. Rosenschein JS, Zlotkin G (1994) Rules of encounter: designing conventions for automated negotiation among computers. MIT Press, Cambridge
16. Lomuscio AR, Wooldridge MJ, Jennings NR (2003) A classification scheme for negotiation in electronic commerce. Group Decis Negot 12(1):31–56

17. Marsa-Maestre I, Klein M, Jonker CM, Reyhan A (2013) From problems to protocols: towards a negotiation handbook. Decis Support Syst
18. Osborne MJ, Rubinstein A (1994) A course in game theory, 1st edn. MIT Press, Cambridge
19. Rubinstein A (1982) Perfect equilibrium in a bargaining model. Econometrica 50(1):97–109
20. Fatima SS, Wooldridge MJ, Jennings NR (2002) Multi-issue negotiation under time constraints. AAMAS'02: Proceedings of the first international joint conference on autonomous agents and multiagent systems. ACM, New York, pp 143–150
21. Kraus S, Wilkenfeld J, Zlotkin G (1995) Multiagent negotiation under time constraints. Artif Intell 75(2):297–345
22. Osborne MJ, Rubinstein A (1990) Bargaining and markets (Economic theory, econometrics, and mathematical economics). Academic Press, London
23. Raiffa H (1982) The art and science of negotiation: How to resolve conflicts and get the best out of bargaining. Harvard University Press, Cambridge
24. Sofer I, Sarne D, Hassidim A (2012) Negotiation in exploration-based environment. In: Proceedings of the twenty-sixth AAAI conference on artificial intelligence
25. Fatima SS, Wooldridge MJ, Jennings NR (2002) Optimal negotiation strategies for agents with incomplete information. Revised papers from the 8th international workshop on intelligent agents VIII, ATAL'01. Springer, London, pp 377–392
26. Sierra C, Faratin P, Jennings NR (1997) A service-oriented negotiation model between autonomous agents. In: Boman M, van de Velde W (eds) Proceedings of the 8th European workshop on modelling autonomous agents in multi-agent world, MAAMAW-97. Lecture notes in artificial intelligence, vol 1237. Springer, Heidelberg, pp 17–35
27. Faratin P, Sierra C, Jennings NR (1998) Negotiation decision functions for autonomous agents. Robot Auton Syst 24(3–4):159–182
28. Niemann C, Lang F (2009) Assess your opponent: A bayesian process for preference observation in multi-attribute negotiations. In: Ito T, Zhang M, Robu V, Fatima S, Matsuo T (eds) Advances in agent-based complex automated negotiations, Studies in computational intelligence, vol 233. Springer, Berlin, pp 119–137
29. Fatima SS, Wooldridge MJ, Jennings NR (2006) Multi-issue negotiation with deadlines. J Artif Intell Res 27:381–417
30. Slembeck T (1999) Reputations and fairness in bargaining-experimental evidence from arepeated ultimatum game with fixed opponents. Experimental, EconWPA
31. Santiago Sánchez-Pagés. The use of conflict as a bargaining tool against unsophisticated opponents. ESE Discussion Papers 99, Edinburgh School of Economics, University of Edinburgh, 2004
32. An B, Gatti N, Lesser VR (2013) Bilateral bargaining with one-sided uncertain reserve prices. Auton Agents Multi-Agent Syst 26:420–455
33. Chevaleyre Y, Dunne PE, Endriss U, Lang J, Lemaître M, Maudet N, Padget J, Phelps S, Rodríguez-Aguilar JA, Sousa P (2006) Issues in multiagent resource allocation. Informatica 30:3–31
34. Domshlak C, Hüllermeier E, Kaci S, Prade H (2011) Preferences in AI: an overview. Artif Intell 175(7–8):1037–1052
35. Aydoğan R, Baarslag T, Hindriks KV, Jonker CM, Yolum P (2013) Heuristic-based approaches for CP-nets in negotiation. In: Ito T, Zhang M, Robu V, Matsuo T (eds) Complex automated negotiations: Theories, models, and software competitions, Studies in computational intelligence, vol 435. Springer, Berlin, pp 113–123
36. Aydoğan R, Baarslag T, Hindriks KV, Jonker CM, Yolum P (2014) Heuristics for using CP-nets in utility-based negotiation without knowing utilities. Knowl Inf Syst 1–32
37. Boutilier C, Brafman RI, Domshlak C, Hoos HH, Poole D (2004) CP-nets: A tool for representing and reasoning with conditionalceteris paribus preference statements. J Artif Intell Res 21:135–191
38. Keeney RL, Raiffa H (1976) Decisions with multiple objectives. Cambridge University Press, Cambridge

39. Raiffa H, Richardson J, Metcalfe D (2003) Negotiation analysis: The science and art of collaborative decision making. Harvard University Press, Cambridge
40. Ito T, Zhang M, Robu V, Fatima S, Matsuo T (2011) New trends in agent-based complex automated negotiations, vol 383. Springer Science & Business Media, New York
41. Marsa-Maestre I, Lopez-Carmona MA, Ito T, Zhang M, Bai Q, Fujita K (2014) Novel insights in agent-based complex automated negotiation, vol 535. Springer, Japan
42. de Jonge D (2015) Negotiations over large agreement spaces. Ph.D. thesis, Universitat Autònoma de Barcelona
43. Ito T, Klein M, Hattori H (2008) A multi-issue negotiation protocol among agents with nonlinearutility functions. Multiagent Grid Syst 4(1):67–83
44. Klein M, Faratin P, Sayama H, Bar-Yam Y (2003) Negotiating complex contracts. Group Decis Negot 12:111–125
45. Lopez-Carmona MA, Marsa-Maestre I, Klein M, Ito T (2012) Addressing stability issues in mediated complex contract negotiations for constraint-based, non-monotonic utility spaces. Auton Agents Multi-Agent Syst 24(3):485–535
46. Robu V, Somefun K, La Poutré JA (2005) Modeling complex multi-issue negotiations using utility graphs. Proceedings of the fourth international joint conference on autonomous agents and multiagent systems, AAMAS'05. ACM, New York, pp 280–287
47. Marsa-Maestre I, Lopez-Carmona MA, Velasco JR, Ito T, Klein V, Fujita K (2009) Balancing utility and deal probability for auction-based negotiations in highly nonlinear utility spaces. In: Proceedings of the 21st international joint conference on artifical intelligence, IJCAI'09. Morgan Kaufmann Publishers Inc., pp 214–219
48. Hadfi R, Ito T (2014) Addressing complexity in multi-issue negotiation via utility hypergraphs. In: Proceedings of the twenty-eighth AAAI conference on artificial intelligence
49. Chevaleyre Y, Endriss U, Estivie S, Maudet N (2004) Multiagent resource allocation with k-additive utility functions. In: Proceedings of the DIMACS-LAMSADE workshop on computerscience and decision theory, pp 83–100
50. Nisan N (2006) Bidding languages. Combinatorial auctions, Cambridge
51. Zeng D, Sycara KP (1997) Benefits of learning in negotiation. In: Proceedings of the fourteenth national conference on artificial intelligence and ninth conference on innovative applications of artificial intelligence, AAAI'97/IAAI'97. AAAI Press, pp 36–41
52. Zeng D, Sycara KP (1998) Bayesian learning in negotiation. Int J Hum Comput Stud 48(1):125–141
53. Carnevale PJD, Lawler EJ (1986) Time pressure and the development of integrative agreements in bilateral negotiations. J Conflict Resolut 30(4):636–659
54. Pruitt DG (1981) Negotiation behavior. Academic Press, New York
55. Kersten GE, Noronha SJ (1998) Rational agents, contract curves, and inefficient compromises. Trans Syst Man Cybern Part A 28(3):326–338
56. Hindriks KV, Tykhonov D (2010) Towards a quality assessment method for learning preference profiles in negotiation. In: Ketter W, La Poutré JA, Sadeh N, Shehory O, Walsh W (eds) Agent-Mediated electronic commerce and trading agent design and analysis. Lecture notes in business information processing, vol 44. Springer, Berlin, pp 46–59
57. Cheng C-B, Chan C-CH, Lin K-C (2006) Intelligent agents for e-marketplace: Negotiation with issue trade-offs by fuzzy inference systems. Decis Support Syst 42(2):626–638
58. Faratin P, Sierra C, Jennings NR (2002) Using similarity criteria to make issue trade-offs in automated negotiations. Artif Intell 142(2):205–237
59. Ito T, Hattori H, Klein M (2007) Multi-issue negotiation protocol for agents: exploring nonlinear utility spaces. Proceedings of the 20th international joint conference on artifical intelligence, IJCAI'07. Morgan Kaufmann Publishers Inc, San Francisco, pp 1347–1352
60. Jonker CM, Robu V, Treur J (2007) An agent architecture for multi-attribute negotiation using incomplete preference information. Auton Agents Multi-Agent Syst 15:221–252
61. Lin R, Oshrat Y, Kraus S (2009) Investigating the benefits of automated negotiations in enhancing people's negotiation skills. In: AAMAS'09: Proceedings of the 8th international conference on autonomous agents and multiagent systems. International foundation for autonomous agents and multiagent systems, Richland, pp 345–352

62. Karp AH, Wu R, Chen K-Y, Zhang A (2004) A game tree strategy for automated negotiation. In: Proceedings of the 5th ACM conference on electronic commerce, EC'04. ACM, New York, pp 228–229

63. Lin R, Kraus S, Wilkenfeld J, Barry J (2008) Negotiating with bounded rational agents in environments with incomplete information using an automated agent. Artif Intell 172(6–7):823–851

64. Kawaguchi S, Fujita K, Ito T (2012) Compromising strategy based on estimated maximum utility for automated negotiating agents. In: Ito T, Zhang M, Robu V, Fatima S, Matsuo T (eds) New trends in agent-based complex automated negotiations. Series of studies in computational intelligence, Springer, Berlin, pp 137–144

65. Kawaguchi S, Fujita K, Ito T (2013) AgentK2: Compromising strategy based on estimated maximum utility for automated negotiating agents. In: Ito T, Zhang M, Robu V, Matsuo T (eds) Complex automated negotiations: Theories, models, and software competitions. Studies in computational intelligence, vol 435. Springer, Berlin, pp 235–241

66. Williams CR (2012) Practical strategies for agent-based negotiation in complex environments. Ph.D. thesis, University of Southampton, 2012

67. Williams CR, Robu V, Gerding EH, Jennings NR (2012) Iamhaggler: A negotiation agent for complex environments. In: Ito T, Zhang M, Robu V, Fatima S, Matsuo T (eds) New trends in agent-based complex automated negotiations, Studies in computational intelligence. Springer, Berlin, pp 151–158

68. Williams CR, Robu V, Gerding EH, Jennings NR (2013) Iamhaggler 2011: A gaussian process regression based negotiation agent. In: Ito T, Zhang M, Robu V, Matsuo T (eds) Complex automated negotiations: theories, models, and software competitions, Studies in computational intelligence, vol 435. Springer, Berlin, pp 209–212

69. Hao J, Leung H-F (2012) ABiNeS: An adaptive bilateral negotiating strategy over multiple items. Proceedings of the The 2012 IEEE/WIC/ACM international joint conferences on web intelligence and intelligent agent technology, WI-IAT'12, vol 2. IEEE Computer Society, Washington, pp 95–102

70. Hao J, Leung H-F (2014) CUHK agent: An adaptive negotiation strategy for bilateral negotiations over multiple items. In: Marsa-Maestre I, Lopez-Carmona MA, Ito T, Zhang M, Bai Q, Fujita K (eds) Novel insights in agent-based complex automated negotiation, Studies in computational intelligence, vol 535. Springer, Japan, pp 171–179

71. Chen S, Ammar HB, Tuyls K, Weiss G (2013) Optimizing complex automated negotiation using sparse pseudo-input gaussian processes. In: Proceedings of the 2013 international conference on autonomous agents and multi-agent systems, AAMAS'13. International foundation for autonomous agents and multiagent systems, Richland, pp 704–714

72. Chen S, Weiss G (2012) An efficient and adaptive approach to negotiation in complex environments. In: De Raedt L, Bessiere C, Dubois D, Doherty P, Frasconi P, Heintz F, Lucas PJF (eds) ECAI, Frontiers in artificial intelligence and applications, vol 242. IOS Press, pp 228–233

73. Chen S, Weiss G (2012) A novel strategy for efficient negotiation in complex environments. In: Timm IJ, Guttmann C (eds) Multiagent system technologies, Lecture notes in computer science, vol 7598. Springer, Berlin, pp 68–82

74. Chen S, Weiss G (2013) An efficient automated negotiation strategy for complex environments. Eng Appl Artif Intell 26(10)

75. Ilany L, Gal Y(K) (2014) The simple-meta agent. In: Marsa-Maestre I, Lopez-Carmona MA, Ito T, Zhang M, Bai Q, Fujita K (eds) Novel insights in agent-based complex automated negotiation. Studies in computational intelligence, vol 535. Springer, Japan, pp 197–200

76. Ilany L, Gal Y (2015) Algorithm selection in bilateral negotiation. Auton Agents Multi-Agent Syst 1–27

77. Ilany L, Gal Y (2013) Algorithm selection in bilateral negotiation. In: Proceedings of the twenty-seventh AAAI conference on artificial intelligence (AAAI 2013)

78. Bartolini C, Preist C, Jennings NR (2002) A generic software framework for automated negotiation. In: First international conference on autonomous agent and multi-agent systems

79. Dumas M, Governatori G, Ter Hofstede AHM, Oaks P (2002) A formal approach to negotiating agents development. Electron Commer Res Appl 1(2):193–207
80. Ashri R, Rahwan I, Luck M (2003) Architectures for negotiating agents. In: Proceedings of the 3rd central and Eastern European conference on multi-agent systems. Springer, pp 136–146
81. Hindriks KV, Jonker CM, Tykhonov D (2008) Towards an open negotiation architecture for heterogeneous agents. In: Klusch M, Pechoucek M, Polleres A (eds) Cooperative information agents XII. Lecture notes in computer science, vol 5180. Springer, Berlin, pp 264–279
82. Matos N, Sierra C, Jennings NR (1998) Determining successful negotiation strategies: an evolutionary approach. In: Proceedings international conference on multi agent systems, pp 182–189
83. Eymann T (2001) Co-evolution of bargaining strategies in a decentralized multi-agent system. In: AAAI fall 2001 symposium on negotiation methods for autonomous cooperative systems, pp 126–134
84. Beam C, Segev A (1997) Automated negotiations: a survey of the state of the art. Wirtschaftsinformatik 39(3):263–268
85. Ros R, Sierra C (2006) A negotiation meta strategy combining trade-off and concession moves. Auton Agents Multi-Agent Syst 12:163–181
86. Lai H, Doong H-S, Kao C-C, Kersten GE (2006) Negotiators' communication, perception of their counterparts, and performance in dyadic e-negotiations. Group Decis Negot 15:429–447
87. Lomuscio AR, Wooldridge MJ, Jennings NR (2001) A classification scheme for negotiation in electronic commerce. In: Dignum F, Sierra C (eds) Agent mediated electronic commerce, Lecture notes in computer science, vol 1991. Springer, Berlin, pp 19–33
88. Thomas KW (1992) Conflict and conflict management: Reflections and update. J Organ Behav 13(3):265–274
89. Zachariassen F (2008) Negotiation strategies in supply chain management. Int J Phys Distrib Logistics Manage 38:764–781
90. An B, Sim KM, Tang LG, Miao CY, Shen ZQ, Cheng DJ (2008) Negotiation agents' decision making using markov chains. In: Ito T, Hattori H, Zhang M, Matsuo T (eds) Rational, robust, and secure negotiations in multi-agent systems, Studies in computational intelligence, vol 89. Springer, Berlin, pp 3–23
91. Lewicki RJ, Saunders DM, Barry B, Minton JW (2003) Essentials of negotiation. McGraw-Hill, Boston, MA
92. Gode DK, Sunder S (1993) Allocative efficiency in markets with zero intelligence (ZI) traders: Market as a partial substitute for individual rationality. J Polit Econ 101(1):119–137
93. Axelrod R (1984) The evolution of cooperation. Basic Books, New York
94. Axelrod R, Dion D (1998) The further evolution of cooperation. Science 242(4884):1385–1390
95. Baarslag T, Hindriks KV, Jonker CM (2013) A tit for tat negotiation strategy for real-time bilateral negotiations. In: Ito T, Zhang M, Robu V, Matsuo T (eds) Complex automated negotiations: theories, models, and software competitions, Studies in computational intelligence, vol 435. Springer, Berlin, Heidelberg, pp 229–233
96. Hindriks, Jonker CM, Tykhonov D (2009) The benefits of opponent models in negotiation. In: Proceedings of the 2009 IEEE/WIC/ACM international joint conference on web intelligence and intelligent agent technology, vol 2. IEEE Computer Society, pp 439–444
97. Rahwan I, Ramchurn S, Jennings NR, McBurney P, Parsons S, Sonenberg L (2003) Argumentation-based negotiation. Knowl Eng Rev 18(04):343–375
98. Sandholm T, Lesser VR (1996) Advantages of a leveled commitment contracting protocol. In: Clancey WJ, Weld DS (eds) Proceedings of the thirteenth national conference on artificial intelligence and eighth innovative applications of artificial intelligence conference, AAAI 96, IAAI 96, vol 1. AAAI Press/The MIT Press, Portland, Oregon, pp 126–133, 4–8 Aug 1996
99. Kolomvatsos K, Hadjieftymiades S (2014) On the use of particle swarm optimization and kernel density estimator in concurrent negotiations. Inf Sci 262:99–116
100. Williams CR, Robu V, Gerding EH, Jennings NR (2012) Towards a platform for concurrent negotiations in complex domain. In: Proceedings of the fifth international workshop on agent-based complex automated negotiations (ACAN 2012)

101. Williams CR, Robu V, Gerding EH, Jennings NR (2012) Negotiating concurrently with unknown opponents in complex, real-time domains. In: 20th European conference on artificial intelligence, vol 242, pp 834–839

102. Kolomvatsos K, Anagnostopoulos C, Hadjiefthymiades S (2013) Determining the optimal stopping time for automated negotiations. IEEE Trans Syst Man Cybern Syst 99:1–1

103. Leonardz B (1973) To stop or not to stop some elementary optimal stopping problems with economic interpretations. Almqvist & Wiksell, Stockholm

104. Mengxiao W, de Weerdt M, La Poutré JA (2013) Acceptance strategies for maximizing agent profits in online scheduling. In: David E, Robu V, Shehory O, Stein S, Symeonidis A (eds) Agent-Mediated electronic commerce. Designing trading strategies and mechanisms for electronic markets, Lecture notes in business information processing. vol 119. Springer, Berlin, Heidelberg, pp 115–128

105. van den Herik J, Donkers J, Spronck PHM (2005) Opponent modelling and commercial games. In: Kendall G, Lucas S (eds) Proceedings of the IEEE 2005 symposium on computational intelligence and games, pp 15–25

106. Baarslag T, Hendrikx MJC, Hindriks KV, Jonker CM (2015) Learning about the opponent in automated bilateral negotiation: a comprehensive survey of opponent modeling techniques. Auton Agents Multi-Agent Syst pp 1–50

107. Schadd FC, Bakkes S, Spronck PHM (2007) Opponent modeling in real-time strategy games. In: 8th International conference on intelligent games and simulation (GAME-ON 2007), pp 61–68

108. Hindriks KV, Jonker CM, Tykhonov D (2011) Let's dans! An analytic framework of negotiation dynamics and strategies. Web Intell Agent Syst 9(4):319–335

109. Rahman SA, Bahgat R, Farag GM (2011) Order statistics bayesian-mining agent modelling for automated negotiation. Informatica Int J Comput Inf 35(1):123–137

110. Jeonghwan Gwak and Kwang Mong Sim (2011) Bayesian learning based negotiation agents for supporting negotiation with incomplete information. Proceedings of the international multiconference of engineers and computer scientists 1:163–168

111. Haberland V, Miles S, Luck M (2012) Adaptive negotiation for resource intensive tasks in grids. In: STAIRS, pp 125–136

112. Jazayeriy H, Azmi-Murad M, Sulaiman N, Udizir N (2011) The learning of an opponent's approximate preferences in bilateral automated negotiation. J Theor Appl Electron Commer Res 6(3):65–84

113. Klos TB, Somefun K, La Poutré JA (2011) Automated interactive sales processes. IEEE Intell Syst 26(4):54–61

114. Oprea M (2002) An adaptive negotiation model for agent-based electronic commerce. Stud Inform Control 11(3):271–279

115. Oshrat Y, Lin R, Kraus S (2009) Facing the challenge of human-agent negotiations via effective general opponent modeling. Proceedings of the 8th international conference on autonomous agents and multiagent systems, AAMAS '09, vol 1. International foundation for autonomous agents and multiagent systems. Richland, SC, pp 377–384

116. Papaioannou IV, Roussaki IG, Anagnostou ME (2011) Multi-modal opponent behaviour prognosis in e-negotiations. In: Cabestany J, Rojas I, Joya G (eds) Advances in computational intelligence, vol 6691., Lecture notes in computer scienceSpringer, Berlin Heidelberg, pp 113–123

117. Papaioannou IV, Roussaki IG, Anagnostou ME (2008) Neural networks against genetic algorithms for negotiating agent behaviour prediction. Web Intell Agent Syst 6(2):217–233

118. Rau H, Tsai M-H, Chen C-W, Shiang W-J (2006) Learning-based automated negotiation between shipper and forwarder. Comput Ind Eng 51(3):464–481

119. Sim KM, Guo Y, Shi B (2009) BLGAN: Bayesian learning and genetic algorithm for supporting negotiation with incomplete information. IEEE Trans Syst Man Cybern Part B Cybern 39(1):198–211

120. Sim KM, Guo Y, Shi B (2007) Adaptive bargaining agents that negotiate optimally and rapidly. IEEE Congr Evol Comput IEEE, pp 1007–1014

121. Agrawal MK, Chari K (2009) Learning negotiation support systems in competitive negotiations: A study of negotiation behaviours and system impacts. Int J Intell Inf Technol 5(1):1–23
122. Brzostowski J, Kowalczyk R (2006) Predicting partner's behaviour in agent negotiation. Proceedings of the fifth international joint conference on autonomous agents and multiagent systems, AAMAS'06. ACM, New York, NY, USA, pp 355–361
123. Jakub Brzostowski J, Kowalczyk R (2006) Adaptive negotiation with on-line prediction of opponent behaviour in agent-based negotiations. Proceedings of the IEEE/WIC/ACM international conference on intelligent agent technology, IAT'06. IEEE Computer Society, Washington, DC, USA, pp 263–269
124. Fang F, Xin Y, Yun X, Haitao X (2008) An opponent's negotiation behavior model to facilitate buyer-seller negotiations in supply chain management. In: Electronic commerce security, international symposium, pp 582–587
125. Hou C (2004) Predicting agents tactics in automated negotiation. In: Proceedings of the IEEE/WIC/ACM international conference on intelligent agent technology. IEEE Computer Society, pp 127–133
126. Masvoula M, Halatsis C, Martakos D (2011) Predictive automated negotiators employing risk-seeking and risk-averse strategies. In: Iliadis L, Jayne C (eds) Engineering applications of neural network. IFIP Advances in information and communication technology, vol 363. Springer Boston, pp 325–334
127. Mudgal C, Vassileva J (2000) Bilateral negotiation with incomplete and uncertain information: A decision-theoretic approach using a model of the opponent. Proceedings of the 4th international workshop on cooperative information agents IV, the future of information agents in cyberspace, CIA'00. Springer, London, UK, pp 107–118
128. Narayanan V, Jennings NR (2006) Learning to negotiate optimally in non-stationary environments. In: Klusch M, Rovatsos M, Payne TR (eds) Cooperative information agents X, vol 4149., Lecture notes in computer scienceSpringer, Berlin, pp 288–300
129. Ozonat K, Singhal S (2010) Design of negotiation agents based on behavior models. In: Chen L, Triantafillou P, Suel T (eds) Web information systems engineering-WISE 2010, vol 6488., Lecture notes in computer scienceSpringer, Berlin, pp 308–321
130. Ren Z, Anumba CJ (2002) Learning in multi-agent systems: A case study of construction claims negotiation. Adv Eng Inf 16(4):265–275
131. Saha S, Biswas A, Sen S (2005) Modeling opponent decision in repeated one-shot negotiations. Proceedings of the fourth international joint conference on autonomous agents and multiagent systems, AAMAS'05. ACM, New York, NY, USA, pp 397–403
132. Williams CR, Robu V, Gerding EH, Jennings NR (2011) Using gaussian processes to optimise concession in complex negotiations against unknown opponents. In: Proceedings of the twenty-second international joint conference on artificial intelligence, IJCAI'11, vol 1. AAAI Press, pp 432–438
133. Chao Y, Ren F, Zhang M (2013) An adaptive bilateral negotiation model based on Bayesian learning. In: Ito T, Zhang M, Robu V, Matsuo T (eds) Complex automated negotiations: Theories, models, and software competitions, vol 435., Studies in computational intelligenceSpringer, Berlin Heidelberg, pp 75–93
134. Fatima S, Wooldridge MJ, Jennings NR (2004) Optimal negotiation of multiple issues in incomplete information settings. In: Proceedings of the third international joint conference on autonomous agents and multiagent systems, vol 3. IEEE Computer Society, pp 1080–1087
135. Baarslag T, Hendrikx MJC, Hindriks KV, Jonker CM (2012) Measuring the performance of online opponent models in automated bilateral negotiation. In: Thielscher M, Zhang D (eds) AI 2012: Advances in artificial intelligence, vol 7691., Lecture notes in computer scienceSpringer, Berlin, pp 1–14
136. Coehoorn RM, Jennings NR (2004) Learning an opponent's preferences to make effective multi-issue negotiation trade-offs. Proceedings of the 6th international conference on electronic commerce, ICEC'04. ACM, New York, NY, USA, pp 59–68
137. Lau RYK, Li Y, Song D, Kwok RC-W (2008) Knowledge discovery for adaptive negotiation agents in e-marketplaces. Decis Support Syst 45(2):310–323

138. Lin R, Kraus S, Wilkenfeld J, Barry J (2006) An automated agent for bilateral negotiation with bounded rational agents with incomplete information. Proceedings of the 2006 conference on ECAI 2006: 17th European conference on artificial intelligence. IOS Press, Amsterdam, The Netherlands, pp 270–274

139. Mok WWH, Sundarraj RP (2005) Learning algorithms for single-instance electronic negotiations using the time-dependent behavioral tactic. ACM Trans Internet Technol 5(1):195–230

140. Somefun K, La Poutré JA (2007) A fast method for learning non-linear preferences online using anonymous negotiation data. In: Fasli M, Shehory O (eds) Agent-Mediated electronic commerce. Automated negotiation and strategy design for electronic markets, Lecture notes in computer science, vol 4452. Springer, Berlin, pp 118–131

141. Zhang M, Tan Z, Zhao J, Li L (2008) A bayesian learning model in the agent-based bilateral negotiation between the coal producers and electric power generators. In: International symposium on intelligent information technology application workshops, IITAW '08, pp 859–862, Dec 2008

142. Baarslag T, Hendrikx MJC, Hindriks KV, Jonker CM (2013) Predicting the performance of opponent models in automated negotiation. In: International joint conferences on web intelligence (WI) and intelligent agent technologies (IAT), 2013 IEEE/WIC/ACM, vol 2, pp 59–66, Nov 2013

143. Bosse T, Jonker CM, van der Meij L, Robu V, Treur J (2005) A system for analysis of multi-issue negotiation. In: Unland R, Calisti M, Klusch M (eds) Software agent-based applications, platforms and development kits, whitestein series in software agent technologies. Birkhöuser Basel, pp 253–279

144. Buffett S, Comeau L, Spencer B, Fleming MW (2006) Detecting opponent concessions in multi-issue automated negotiation. Proceedings of the 8th international conference on electronic commerce: The new e-commerce: Innovations for conquering current barriers, obstacles and limitations to conducting successful business on the internet, ICEC'06. ACM, New York, NY, USA, pp 11–18

145. Frieder A, Miller G (2013) Value model agent: A novel preference profiler for negotiation with agents. In: Ito T, Zhang M, Robu V, Matsuo T (eds) Complex automated negotiations: Theories, models, and software competitions, Studies in computational intelligence, vol 435. Springer, Berlin, pp 199–203

146. Hindriks KV, Tykhonov D (2008) Opponent modelling in automated multi-issue negotiation using bayesian learning. Proceedings of the 7th international joint conference on autonomous agents and multiagent systems, AAMAS'08, vol 1. International foundation for autonomous agents and multiagent systems. Richland, SC, pp 331–338

147. Jonker CM, Robu V (2004) Automated multi-attribute negotiation with efficient use of incomplete preference information. Proceedings of the third international joint conference on autonomous agents and multiagent systems, AAMAS'04, IEEE Computer Society, vol 3. Washington, DC, USA, pp 1054–1061

148. Robu V, La Poutré JA (2006) Retrieving the structure of utility graphs used in multi-item negotiations through collaborative filtering of aggregate buyer preferences. In: Proceedings of the 2nd international workshop on rational, robust and secure negotiations in MAS. Springer

149. Saha S, Sen S (2005) A bayes net approach to argumentation based negotiation. In: Rahwan I, Moraïtis P, Reed C (eds) Argumentation in multi-agent systems, vol 3366., Lecture notes in computer scienceBerlin, Heidelberg, pp 208–222

150. Şerban LD, Silaghi GC, Litan CM (2012) AgentFSEGA-time constrained reasoning model for bilateral multi-issue negotiations. In: Ito T, Zhang M, Robu V, Fatima S, Matsuo T (eds) New trends in agent-based complex automated negotiations. Series of studies in computational intelligence, Springer, Berlin, Heidelberg, pp 159–165

151. van Galen Last N (2012) Agent Smith: Opponent model estimation in bilateral multi-issue negotiation. In: Ito T, Zhang M, Robu V, Fatima S, Matsuo T (eds) New trends in agent-based complex automated negotiations. Studies in computational intelligence, Springer, Berlin, Heidelberg, pp 167–174

152. van Krimpen T, Looije D, Hajizadeh S (2013) Hardheaded. In: Ito T, Zhang M, Robu V, Matsuo T (eds) Complex automated negotiations: Theories, models, and software competitions, Studies in computational intelligence, vol 435. Springer, Berlin, pp 223–227

153. Aydoğan R, Yolum P (2012) The effect of preference representation on learning preferences in negotiation. In: Ito T, Zhang M, Robu V, Fatima S, Matsuo T (eds) New trends in agent-based complex automated negotiations, Studies in computational intelligence, vol 383., Springer-Berlin, Heidelberg, pp 3–20

154. Aydoğan R, Yolum P (2006) Learning consumer preferences for content-oriented negotiation. In: AAMAS Workshop on business agents and the semantic web (BASeWEB). ACM, Press, pp 43–52, May 2006

155. Aydoğan R, Yolum P (2012) Learning opponent's preferences for effective negotiation: An approach based on concept learning. Auton Agents Multi-Agent Syst 24:104–140

156. Farag GM, AbdelRahman SES, Bahgat R, A-Moneim Atef M. Towards KDE mining approach for multi-agent negotiation. In: 2010 The 7th International conference on informatics and systems (INFOS). IEEE, pp 1–7, Mar 2010

157. Buffett S, Spencer B (2005) Learning opponents' preferences in multi-object automated negotiation. Proceedings of the 7th international conference on electronic commerce, ICEC '05. ACM, New York, NY, USA, pp 300–305

158. Buffett S, Spencer B (2007) A bayesian classifier for learning opponents' preferences in multi-object automated negotiation. Electron Commer Res Appl 6(3):274–284

159. Restificar A, Haddawy P (2004) Inferring implicit preferences from negotiation actions. In: International symposium on artificial intelligence and mathematics. Fort Lauderdale, Florida, USA, Jan 2004

160. Ren F, Zhang M (2007) Predicting partners' behaviors in negotiation by using regression analysis. In: Zhang Z, Siekmann J (eds) Knowledge science, engineering and management, vol 4798., Lecture notes in computer scienceSpringer, Berlin, Heidelberg, pp 165–176

161. Masvoula M (2013) Forecasting negotiation counterpart's offers: A focus on session-long learning agents. In: COGNITIVE 2013, the fifth international conference on advanced cognitive technologies and applications, pp 71–76

162. Carbonneau RA, Kersten GE, Vahidov RM (2008) Predicting opponent's moves in electronic negotiations using neural networks. Expert Syst Appl 34(2):1266–1273

163. Carbonneau RA, Kersten GE, Vahidov RM (2011) Pairwise issue modeling for negotiation counteroffer prediction using neural networks. Decis Support Syst 50(2):449–459

164. Lee CC, Ou-Yang C (2009) A neural networks approach for forecasting the supplier's bid prices in supplier selection negotiation process. Expert Syst Appl 36(2, Part 2):2961–2970

165. Lin R, Kraus S, Baarslag T, Tykhonov D, Hindriks KV, Jonker CM (2014) Genius: An integrated environment for supporting the design of generic automated negotiators. Comput Intell 30(1):48–70

166. Devaux L, Paraschiv C (2001) Bargaining on an internet agent-based market: Behavioral vs. optimizing agents. Electron Commer Res 1:371–401

167. Fatima SS, Wooldridge MJ, Jennings NR (2005) A comparative study of game theoretic and evolutionary models of bargaining for software agents. Artif Intell Rev 23(2):187–205

168. Henderson P, Crouch S, Walters RJ, Ni Q (2003) Comparison of some negotiation algorithms using a tournament-based approach. In: Carbonell JG, Siekmann J, Kowalczyk R, Müller JP, Tianfield H, Unland R (eds) Agent technologies, infrastructures, tools, and applications for E-Services, vol 2592. Lecture notes in computer science. springerBerlin, Heidelberg, pp 137–150

169. Manistersky E, Lin R, Kraus S (2008) Understanding how people design trading agents over time. In: Proceedings of AAMAS'08, pp. 1593–1596

170. Hindriks KV, Jonker CM, Kraus S, Lin R, Tykhonov D (2009) Genius: negotiation environment for heterogeneous agents. Proceedings of the 8th international conference on autonomous agents and multiagent systems, AAMAS'09, vol 2. International foundation for autonomous agents and multiagent systems. Richland, SC, pp 1397–1398

171. Lin R, Kraus S, Tykhonov D, Hindriks KV, Jonker CM (2011) Supporting the design of general automated negotiators. In: Proceedings of the second international workshop on agent-based complex automated negotiations (ACAN'09), vol 319. Springer, pp 69–87

172. Thiessen EM, Loucks DP, Stedinger JR (1998) Computer-assisted negotiations of water resources conflicts. GDN 7(2):109–129

173. Kersten GE, Noronha SJ (1999) WWW-based negotiation support: Design, implementation, and use. Decis Support Syst 25(2):135–154

174. Hales D (2002) Neg-o-net-a negotiation simulation test-bed. Technical Report CPM-02-109, CPM

175. Marsa-Maestre I, Klein M, de la Hoz E, Lopez-Carmona MA (2011) Negowiki: A set of community tools for the consistent comparison of negotiation approaches. In: Kinny D, Hsu JYJ, Governatori G, Ghose AK (eds) Agents in principle, agents in practice. Lecture notes in computer science, vol 7047. Springer, Berlin, Heidelberg, 424–435

176. Littman M, Zinkevich M (2006) The 2006 AAAI computer poker competition. ICGA J 29(3):166

177. Wellman MP, Wurman PR, O'Malley K, Bangera R, de Lin S, Reeves D, Walsh WE (2001) Designing the market game for a trading agent competition. IEEE Internet Comput 5(2):43–51

178. Greenwald A, Stone P (2001) Autonomous bidding agents in the trading agent competition. IEEE Internet Comput 5(2):52–60

179. Ketter W, Collins J, Reddy P, Flath C, de Weerdt M (2011) The power trading agent competition. ERIM report series reference No. ERS-2011-027-LIS

180. Niu J, Cai K, Parsons S, McBurney P, Gerding EH (2010) What the 2007 tac market design game tells us about effective auction mechanisms. Auton Agents Multi-Agent Syst 21:172–203

181. Stone P, Greenwald A (2005) The first international trading agent competition: autonomous bidding agents. Electron Commer Res 5(2):229–265

182. Da Costa AD, Lucena CJ, Da Silva VT, Azevedo SC, Soares FA (2008) Art competition: agent designs to handle negotiation challenges. Trust in agent societies: 11th international workshop, TRUST 2008, Estoril, Portugal, May 12–13, 2008. Revised selected and invited papers. Springer, Berlin, pp 244–272

183. Fullam KK, Klos TB, Muller G, Sabater J, Schlosser A, Barber KS, Rosenschein JS, Vercouter L, Voss M (2005) A specification of the agent reputation and trust (art) testbed: experimentation and competition for trust in agent societies. In: The 4th International joint conference on autonomous agents and multi-agent systems (AAMAS). ACM Press, pp 512–518

184. Baarslag T, Fujita K, Gerding EH, Hindriks KV, Ito T, Jennings NR, Jonker CM, Kraus S, Lin R, Robu V, Williams CR (2013) Evaluating practical negotiating agents: Results and analysis of the 2011 international competition. Artif Intell 198:73–103

185. Baarslag T, Hindriks KV, Jonker CM, Kraus S, Lin R (2012) The first automated negotiating agents competition (ANAC 2010). In: Ito T, Zhang M, Robu V, Fatima S, Matsuo T (eds) New trends in agent-based complex automated negotiations. Studies in computational intelligence, vol 383. Springer, Berlin, Heidelberg, pp 113–135

186. Baarslag T, Hindriks KV, Hendrikx MJC, Dirkzwager ASY, Jonker CM (2014) Decoupling negotiating agents to explore the space of negotiation strategies. In: Marsa-Maestre I, Lopez-Carmona MA, Ito T, Zhang M, Bai Q, Fujita K (eds) Novel insights in agent-based complex automated negotiation. Studies in computational intelligence, vol 535. Springer, Japan, pp 61–83

187. Baarslag T, Hindriks KV, Jonker CM (2011) Towards a quantitative concession-based classification method of negotiation strategies. In: Kinny D, Yung-jen Hsu J, Governatori G, Ghose AK (eds) Agents in principle, agents in practice, lecture notes in computer science, vol 7047. Springer, Berlin, Heidelberg, pp 143–158

188. Baarslag T, Hindriks KV (2013) Accepting optimally in automated negotiation with incomplete information. In: Proceedings of the 2013 international conference on autonomous agents and multi-agent systems. AAMAS '13, Richland. International foundation for autonomous agents and multiagent systems, pp 715–722

Chapter 3
A Component-Based Architecture to Explore the Space of Negotiation Strategies

Abstract In order to study the performance of the *individual components* of a negotiation strategy, we introduce an architecture that distinguishes three components which together constitute a negotiation strategy: *the bidding strategy* (B), *the opponent model* (O), and *the acceptance strategy* (A). When decoupled, the components of different strategies can be recombined to create new strategies. This then allows to pinpoint additional structure in most agent designs and to explore the space of automated negotiating agents. In order to study the performance of the *individual components* of a negotiation strategy, we introduce an architecture that distinguishes three components which together constitute a negotiation strategy: *the bidding strategy* (B), *the opponent model* (O), and *the acceptance strategy* (A). When decoupled, the components of different strategies can be recombined to create new strategies. This then allows to pinpoint additional structure in most agent designs and to explore the space of automated negotiating agents. We implemented our BOA architecture in a generic evaluation environment for negotiating agents (Appendix A), and we amended it with the strategy components of the International Automated Negotiating Agents Competition (Appendix B). In doing so, we have a rich evaluation tool at our disposal, together with a repository that contains many negotiating agents and scenarios. The contribution of this chapter is threefold: first, we show that *existing state-of-the-art agents* are *compatible* with this architecture by re-implementing them in the new framework; second, as an application of our architecture, we systematically *explore the space of possible strategies* by recombining different strategy components, resulting in negotiation strategies that improve upon the current state-of-the-art in automated negotiation; finally, we show how the BOA architecture can be applied to *evaluate the performance of strategy components* and create novel negotiation strategies that outperform the state of the art.

3.1 Introduction

In recent years, many new automated negotiation agents have been developed in the search for an effective, generic automated negotiator. There is now a large body of negotiation strategies available, and with the emergence of the International

© Springer International Publishing Switzerland 2016
T. Baarslag, *Exploring the Strategy Space of Negotiating Agents*,
Springer Theses, DOI 10.1007/978-3-319-28243-5_3

Automated Negotiating Agents Competition (ANAC, see Appendix B), new strategies are generated on a yearly basis.

While methods exist to determine the best negotiation agent given a set of agents (cf. Sect. 2.4), we still do not know which type of agent is most effective in general, and especially why. It is impossible to exhaustively search the large (in fact, infinite) space of negotiation strategies; therefore, there is a need for a systematic way of searching this space for effective candidates.

Many of the sophisticated agent strategies that currently exist are comprised of a fixed set of modules. Generally, a distinction can be made between three different modules: one module that decides whether the opponent's bid is acceptable; one that decides what set of bids could be proposed next; and finally, one that tries to guess the opponent's preferences and takes this into account when selecting an offer to send out. The negotiation strategy is a result of the complex interaction between these components, of which the individual performance may vary significantly. For instance, an agent may contain a module that predicts the opponent's preferences very well, but utility-wise, the agent may still perform badly because it concedes far too quickly.

This entails that overall performance measures, such as average utility obtained in a tournament, make it hard to pinpoint which components of an agent work well. To date, no efficient method exists to identify to which of the components the success of a negotiating agent can be attributed. Finding such a method would allow to develop better negotiation strategies, resulting in better agreements; the idea being that well-performing components together will constitute a well-performing agent.

To tackle this problem, we propose to analyze three components of the agent design separately. We show that most of the currently existing negotiating agents can be fitted into the so-called *BOA architecture* by putting together three main components in a particular way; namely: a *Bidding strategy*, an *Opponent model*, and an *Acceptance condition*. We support this claim by re-implementing, among others, the ANAC agents to fit into our architecture. Furthermore, we show that the BOA agents are equivalent to their original counterparts.

The advantages of fitting agents into the BOA architecture are threefold: first, it allows the study of the behavior and performance of the individual components; second, it allows to systematically explore the space of possible negotiation strategies; third, the identification of isolated components simplifies the creation of new negotiation strategies.

Finally, we demonstrate the value of our architecture by assembling, from already existing components, new negotiating agents that perform better than the agents from which they are created. This shows that by recombining the best performing components, the BOA architecture can yield better performing agents.

The remainder of this chapter is organized as follows. In Sect. 3.2, the BOA agent architecture is introduced, and we outline a research agenda on how to employ it. Section 3.3 provides evidence that many of the currently existing agents fit into the BOA architecture, and discusses challenges in decoupling existing negotiation strategies. Finally, in Sect. 3.4 we discuss lessons learned and provide directions on how we will apply the BOA framework in later chapters.

3.2 The BOA Agent Architecture

In the last decade, many different negotiation strategies have been introduced in the pursuit of a versatile and effective automated negotiator (see related work in Sect. 2.3). Despite this diversity, there is some common structure to the overall design of the agents. For example, every agent decides whether the opponent's offer is acceptable, and if not, what offer should be proposed instead. When the agent decides to make a counter-offer, it considers its own utility, but it usually also takes the opponent's utility into account.

Current work often focuses on optimizing the negotiation strategy as a whole. We propose to direct our attention to a component-based approach, especially now that we have access to a large repository of mutually comparable negotiation strategies due to ANAC. This approach has several advantages:

1. Given measures for the effectiveness of the individual components of a negotiation strategy, we are able to pinpoint the most promising components, which gives insight into the reasons for success of the strategy;
2. Focusing on the most effective components helps to systematically search the space of negotiation strategies by recombining them into new strategies.

In this section, we outline the key components of the BOA agent framework and we outline a research agenda on applying it to current agent design.

3.2.1 The BOA Agent

Based on a survey of literature and the implementations of currently existing negotiation agents, we have identified three main components of a general negotiation strategy: a *bidding strategy*, possibly an *opponent model*, and an *acceptance condition* (BOA). The elements of a BOA agent are visualized in Fig. 3.1.

We make a distinction between two types of components: elements that are part of the agent's *negotiation environment*, and components that are part of the agent itself. The negotiation environment includes the *bidding history* of the ongoing negotiation, the negotiation *domain*, which holds the information of possible bids and other

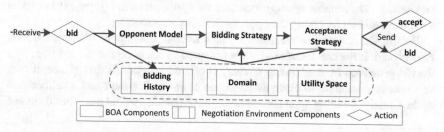

Fig. 3.1 The BOA architecture negotiation flow

negotiation constraints, and the preferences described by the *utility spaces* of the agents (all of which are defined in Chap. 2).

In order to fit an agent into the BOA architecture, it should be possible to distinguish these components in the agent design[1]:

1. **Bidding strategy (B).** At each turn, the *bidding strategy* determines the counter offer by first generating a set of bids, depending on factors such as the opponent's offers, a target threshold, time, and so on. Note that during this stage, the agent only considers what concessions it deems appropriate given its own preferences. The bidding strategy can consult the opponent model (if present) by passing one or multiple bids to see how they compare within the estimated opponent's utility space.

   Input: *opponent utility of bids, negotiation history.*
   Output: *provisional upcoming bid ω.*

2. **Opponent model (O).** An *opponent model* is a learning technique that constructs a model of the opponent's preferences. In our approach, the opponent model should be able to estimate the opponent's utility of any given bid. The BOA architecture focuses on opponent models which estimate the (partial) preference profile, because most existing available implementations fit in this category; however, in principle, our architecture can accommodate for the other types of opponent models as well and may use the preference model to learn other attributes as well (e.g., predicting the opponent's strategy).

   Input: *set of bids B, negotiation history.*
   Output: *estimated opponent utility of the bids in B.*

3. **Acceptance Condition (A).** The *acceptance condition* decides whether the opponent's offer should be accepted. If the opponent's bid is not accepted, the bid generated by the bidding strategy is offered instead.

   Input: *provisional upcoming bid ω, negotiation history.*
   Output: *accept, or send out the upcoming bid ω.*

The components interact in the following way (the full process is visualized in Fig. 3.1): when receiving the opponent's bid, the BOA agent first updates the *bidding history* and *opponent model*, maximizing the information known about the environment and opponent.

Given the opponent bid, the *bidding strategy* determines the counter offer by first generating a set of bids with a similar preference for the agent. Note that during this stage, the agent only considers what concessions it deems appropriate given its own preferences. The *bidding strategy* then uses the *opponent model* (if present) to select a bid from this set by taking the opponent's utility into account.

Finally, the *acceptance condition* decides whether the opponent's action should be accepted. If the opponent's bid is not accepted by the acceptance condition, then the bid generated by the bidding strategy is offered instead. At first glance, it may seem counter-intuitive to make this decision *at the end* of the agent's deliberation cycle. Clearly, deciding upon acceptance *at the beginning* would have the advantage

---

[1] An exposition of the agents we fitted into our framework is given in the next section, which will further motivate the choices made below.

Fig. 3.2 The bidding
strategy sets a target utility
range, which is a subset of
all acceptable outcomes.
From these outcomes, the
opponent model selects the
offers that are also good for
the opponent

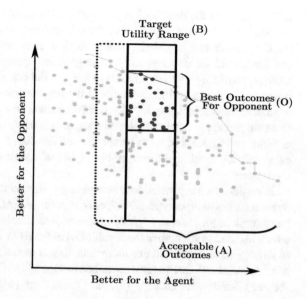

of not wasting resources on generating an offer that might never be sent out. However, generating an offer first allows us to employ acceptance conditions that depend on the utility of the counter bid that is ready to be sent out. This method is widely used in existing agents, as we shall see in our exposition of acceptance conditions in Chap. 4. Such acceptance mechanisms can make a more informed decision by postponing their decision on accepting until the last step; therefore, and given our aim to incorporate as many agent designs as possible, we adopt this approach in our architecture.

To better understand how the different components work together, we might view the negotiation process as a search problem, where the negotiation strategy explores the outcome space for a contract that both parties are willing to agree upon (Fig. 3.2). The bidding strategy controls the rate of concession by setting the *target utility range* (B), which determines the general location of the offer in the outcome space according to the *agent's own utility*. The opponent model can restrict this area even further, by refining the possible offers to bids that are near the Pareto frontier, and hence are *good for the opponent* (O). Finally, the acceptance condition defines the area that consists of all *acceptable outcomes* (A), depending on the jump the agent is willing to make towards the opponent in order to reach an agreement.

3.2.2 Employing the BOA Architecture

We have implemented the BOA architecture as an extension of the GENIUS framework [1] that we outline in Appendix A. The framework was developed as a research tool to facilitate the design of negotiation strategies and to aid in the evaluation of negotiation algorithms. It provides a flexible and easy to use environment for implementing agents and negotiation strategies as well as running negotiations. GENIUS can further aid the development of a negotiation agents by acting as an analytical toolbox, providing a variety of tools to analyze the negotiation agents performance, based on the outcome and dynamics of the negotiation. The BOA architecture has been integrated seamlessly into the GENIUS framework, offering the user the ability to create and apply newly developed components using a graphical user interface as depicted in Fig. 3.3. From the perspective of GENIUS, a negotiation agent is identical to a BOA agent, and therefore both types of agents can participate in the same tournament.

In addition, we organized four annual negotiation competitions (ANAC) that had more than 60 international participants in total. ANAC makes a wide variety of benchmark negotiation strategies and scenarios available to the research community, which we used to strengthen the capabilities of the BOA architecture. The repository of strategies currently contains more than 40 automated negotiation strategies, such as all ANAC 2010–2013 agents described in appendices C–F, the *ABMP strategy* [2], the *Zero Intelligence strategy* [3], the *QO-strategy* [4], the *Bayesian strategy* [5],

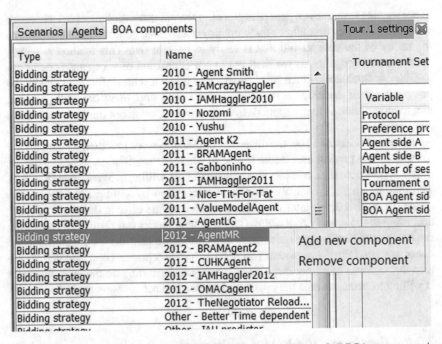

Fig. 3.3 The BOA components window in GENIUS gives an overview of all BOA components in the repository. New components can be added and removed using a graphical user interface. All components can be combined to create new negotiation agents, which then can be evaluated in the analytical toolbox of GENIUS

and others. The repositories of domains and of agents allows us to test the agents on the different domains and against different kinds of strategies.

The framework enables us to follow at least two approaches: first of all, it allows us to independently analyze the components of every negotiation strategy that fits in to our architecture. For example, by re-implementing the ANAC agents in the BOA architecture, it becomes possible to compare the accuracy of all ANAC opponent models, and to pinpoint the best opponent model among them. Following this approach, we are able to identify categories of opponent models that outperform others; naturally, this helps to build better agents in the future.

Secondly, we can proceed to *mix* different BOA components, e.g.: replace the opponent model of the runner-up of ANAC by a different opponent model and then examine whether this makes a difference in placement. Such a procedure enables us to assess the reasons for an agent's success, and makes it possible to systematically search for an effective automated negotiator.

The first part of the approach gives insight in what components are best in isolation; the second part gives us understanding of their influence on the agent as a whole. At the same time, both approaches raise some key theoretical questions, such as:

1. Can the BOA components be identified in all, or at least most, current negotiating agents?
2. How do we measure the performance of the components? Can a single best component be identified, or does this strongly depend on the other components?
3. If the individual components perform better than others (with respect to some performance measure), does combining them in an agent also improve the agent's performance?

In this chapter we do not aim to fully answer all of the above questions; instead, we outline a research agenda for the rest of this thesis, and introduce the BOA architecture as a tool that can be used towards answering these questions.

Nonetheless, in the next section, we will provide empirical support for an affirmative answer to the first theoretical question: indeed, in many cases the components of the BOA architecture can be identified in current agents, and we will also provide reasons for when this is not the case.

The answer to the second question depends on the component under consideration: for an opponent model, it is straightforward to measure its effectiveness [6] (and we will do so in Chaps. 6 and 7): the closer the opponent model is to the actual profile of the opponent, the better it is. The performance of the other two components of the BOA architecture is better measured in terms of utility obtained in negotiation (as will do for acceptance strategies in Chaps. 4 and 5 and for bidding strategies in Chap. 8), as there seems no clear alternative method to define the effectiveness of the acceptance condition or bidding strategy in isolation. In any case, the BOA architecture can be used as a research tool to help answer such theoretical questions.

Regarding the third question: suppose we take the best performing bidding strategy, equip it with the most faithful opponent model, and combine this with the most effective acceptance condition; it would seem reasonable to assume this combination results in an effective negotiator. We elaborate on this conjecture in Chap. 10.

3.3   Decoupling Existing Agents

In this section we provide empirical evidence that many of the currently existing agents can be decoupled by separating the components of a set of state of the art agents. This section serves three goals: first, we discuss how existing agents can be decoupled into a BOA agent; second, we argue that the BOA architecture design is appropriate, as most agents will turn out to fit in our architecture; third, we discuss and apply a method to determine if the sum of the components—the BOA agent—is equal in behavior to the original agent.

3.3.1   Identifying the Components

In this section we identify the components of 21 negotiating agents, taken from the ANAC competition of 2010–2012 as described in Appendix B. We selected these agents as they represent the state of the art in automated negotiation, having been implemented by various negotiation experts.

Since the agents were not designed with decoupling in mind, all agents had to be re-implemented to be supported by the BOA architecture. Our decoupling methodology was to adapt an agent's algorithm to enable it to switch its components, without changing the agent's functionality. A method call to specific functionality, such as code specifying when to accept, was replaced by a more generic call to the acceptance mechanism, which can then be swapped at will. The contract of the generic calls are defined by the expected input and output of every component, as outlined in Sect. 3.2.1.

As an example to illustrate the components within a strategy, we use *IAMhaggler2011* [7], which finished third in ANAC 2011 (see Appendix D). When it receives an offer, the acceptance condition of *IAMhaggler2011* will only accept bids that have a utility higher than a certain predefined value. After that, *IAMHaggler2011* employs an opponent model to approximate the rate at which the opponent concedes, by inspecting the past offered bids. In order to finally choose a counter bid to offer to the opponent, *IAMhaggler2011* first decides on a target utility, based on its own concession rate, which is determined in such a way that the expected utility is optimized.

The first step in decoupling an agent is to determine which components can be identified. For example, in the ANAC 2010 agent *FSEGA* (Appendix C), an acceptance condition, a bidding strategy, and an opponent model can all be identified. The acceptance condition combines simple, utility-based criteria (defined later in this thesis as AC_{const} and AC_{prev}; see Chap. 4), and can be easily decoupled in our architecture. The opponent model is a variant of the *Bayesian opponent model* [5, 8] (which we define later in Sect. 6.4.2), which is used to optimize the opponent utility of a bid. Since this usage is consistent with our architecture (i.e., the opponent model provides opponent utility information), the model can be replaced by a call to the generic opponent model interface. The final step is to change the bidding strategy to use the generic opponent model and acceptance conditions instead of its own specific implementation. In addition to this, the opponent model and acceptance condition need to be altered to allow the other bidding strategies to use it. Other agents can be decoupled using a similar process.

Unfortunately, some agent implementations contain slight dependencies between different components. These dependencies needed to be resolved to separate the design into singular components. For example, the acceptance condition and bidding strategy of the ANAC 2011 agent *The Negotiator*[2] rely on a shared target utility. In such cases, the agent can be decoupled by introducing Shared Agent State (SAS) classes. A SAS class avoids code duplication, and thus performance loss, by sharing the code between the components. One of the components uses the SAS to calculate the values of the required parameters and saves the results, while the other component simply asks for the saved results instead of repeating the calculation.

Table 3.1 provides an overview of all agents that we re-implemented in our architecture, and more specifically, which components we were able to decouple. In fact, we were able to decouple all ANAC 2010, and most ANAC 2011 and ANAC 2012 agents.

There were two agents (*ValueModelAgent* [9] and *Meta-Agent* [10, 11]) that were not decoupled due to practical reasons, even though theoretically it is possible. The *ValueModelAgent* was not decoupled because there were unusually strong dependencies between its components. Decoupling the strategy would result in computationally heavy components when trying to combine them with other components, making them impractical to use. The ANAC 2012 *Meta-Agent* chooses an offer among 17 agents from the ANAC 2011 qualifying round. This agent was not decoupled because it requires the decoupling of all 17 agents, of which only 8 optimized versions entered the finals.

The *CUHK Agent*, like *ValueModelAgent*, is heavily coupled with multiple variables that are shared between the bidding strategy and acceptance condition. This makes it very hard to decouple and can make components unusable in combination with other components (e.g. variables might not properly be set). However, since *CUHK Agent* was placed first in the ANAC 2012 competition, we decided to decouple its bidding strategy, allowing it to work with other acceptance conditions and opponent models.

Four additional agents were only partially decoupled: *AgentLG, BRAMAgent, BRAMAgent2,* and *Gahbininho*. As is evident from Table 3.1, the only obstacle in decoupling these agents fully is their usage of the opponent model, as it can be employed in many different ways. Some agents, such as *Nice Tit for Tat*, attempt to estimate the Nash point on the Pareto frontier. Other common applications include: ranking a set of bids according to the opponent utility, reciprocating in opponent utility, and extrapolating opponent utility. The generic opponent model interface needs to sufficiently accommodate such requirements from the bidding strategy to make interchangeability possible. For this reason we require the opponent model interface

---

[2] Descriptions of all ANAC 2011 agents can be found in Appendix D.

Table 3.1 Overview of the BOA components found in every agent

ANAC 2010	B	O	A
FSEGA [12]	✓	✓	✓
Agent K [13]	✓	∅	✓
Agent Smith [14]	✓	✓	✓
IAMcrazyHaggler [8]	✓	∅	✓
HardHeaded [15]	✓	✓	✓
CUHK Agent [16, 17]	✓	−	−
IAMhaggler [8]	✓	✓	✓
Nozomi [13]	✓	∅	✓
Yushu [18]	✓	∅	✓
ANAC 2011	B	O	A
Agent K2 [19]	✓	∅	✓
BRAMAgent [20]	✓	−	✓
Gahboninho [21]	✓	−	✓
IAMhaggler2011 [22]	✓	∅	✓
Nice Tit for Tat [23]	✓	✓	✓
The Negotiator [24]	✓	∅	✓
ANAC 2012	B	O	A
AgentLG	✓	∅	✓
AgentMR [25]	✓	∅	✓
BRAMAgent2	✓	−	✓
IAMhagger2012	✓	∅	✓
OMAC Agent [26]	✓	∅	✓
The Negotiator Reloaded	✓	✓	✓

✓: original has component, which can be decoupled. ∅: original has no such component, but it can be added. −: no support for such a component

to be able to produce the estimated opponent utility of an arbitrary negotiation outcome.

With regard to the opponent model, there are three groups of agents: first, there are agents such as *FSEGA* [12], which use an opponent model that can be freely interchanged; second, there are agents such as the ANAC 2010 winner *Agent K* [27], which do not have an opponent model themselves, but can be extended to use one. Such agents typically employ a bidding strategy that first decides upon a specific target utility range, and then picks a *random* bid within that range. These agents can easily be fitted with an opponent model instead, by passing the utility range through the opponent model before sending out the bid. Lastly, there are agents, for example *Gahboninho* and *BRAMAgent*, that use a similarity heuristic which is not compatible with our architecture, as their opponent models do not yield enough information to compute the opponent utility of bids. For these type of agents, we consider the opponent model part of the bidding strategy. *AgentLG* also uses an opponent model which is not compatible with our BOA architecture; however, it has been adopted to be able to use other opponent models.

When decoupling the agents, we can distinguish different classes within each component, except for the bidding strategy component, which varies greatly between different agents. For instance, as we will see in Chaps. 6 and 7, there are only three main types of opponent models being used: *Bayesian models*, *Frequency models*, and *Value models*. Bayesian models are an implementation of a (scalable) model of the opponent preferences that is updated using Bayesian learning [5, 28]. The main characteristic of frequency based models is that they track the frequency of occurrence of issues and values in the opponent's bids and use this information to estimate the opponent's preferences. Value models take this approach a step further and solely focus on the frequency of the issue values. In practice, Bayesian models are computationally intensive, whereas frequency and value models are relatively light-weight.

Similar to the opponent models, most agents use variations and combinations of a small set of acceptance conditions. Specifically, many agents use simple thresholds for deciding when to accept (called AC_{const} in Chap. 4) and linear functions that depend on the utility of the bid under consideration (called $AC_{next}(\alpha, \beta)$ in Chap. 4).

3.3.2 Testing Equivalence of BOA Agents

A BOA agent should behave identically to the agent from which its components are derived. Equivalence can be verified in two ways; first, given the same negotiation environment and the same state, both agents should behave in exactly identical ways; second, the performance in a real time negotiation of both agents should be similar.

3.3.2.1 Identical Behavior Test

Two deterministic agents can be considered equivalent if they perform the same action given the same negotiation trace. There are two main problems in determining equivalence: first, most agents are non-deterministic, as they behave randomly in certain circumstances; for example, when picking from a set of bids of similar utility; second, the default protocol in GENIUS uses real time [1], which is highly influenced by CPU performance. This means that in practice, two runs of the same negotiation are never exactly equivalent.

To be able to run an equivalence test despite agents choosing actions at random, we fixed the seeds of the random functions of the agents. The challenge of working in real time was dealt with by changing the real time deadline to a maximum amount of rounds. Since time does not pass within a round, cpu performance does not play a role.

All agents were evaluated on the ANAC 2011 domains (see Appendix D for a domain analysis). The ANAC 2011 domains vary widely in characteristics: the number of issues ranges from 1 to 8, the size from 3 to 390625 possible outcomes, and the discount from none (1.0) to strong (0.424). Some ANAC 2010 agents, specifically *Agent Smith* and *Yushu*, were not designed for large domains and were therefore run on a subset of these domains.

The *opponent strategies* used in the identical behavior test should satisfy two properties: the opponent strategy should be deterministic, and secondly, the opponent strategy should not be the first to accept, to avoid masking errors in the agent's acceptance condition. Given these two criteria, we used the standard time-dependent tactics [29, 30] described in Sect. 2.3.3 for the opponent bidding strategy. Specifically, we use *Hardliner* ($e = 0$), *Conceder Linear* ($e = 1$), and *Conceder* ($e = 2$). In addition, we use the *Offer Decreasing* agent, which offers the set of all possible bids in decreasing order of utility.

All original and BOA agents were evaluated against these four opponents, using both preference profiles defined on all eight ANAC 2011 domains. Both strategies were run in parallel, making sure that the moves made by both agents were equivalent at each moment. After the experiments were performed, the results indicated that all BOA agents were exactly identical to their original counterparts except for *AgentMR* and *AgentLG*. Both these agents do not have identical behavior with its BOA counterpart because of the order in which the components are called; their implementation requires that they first test if the opponent's bid is acceptable, and then determine the bid to offer. As discussed above, this is exactly the opposite of what the BOA agent does.

3.3.2.2   Similar Performance Test

Two agents can perform the same action given the same input, but may still achieve different results because of differences in their real time performance. When decoupling agents, there is a trade-off between the performance and interchangeability of components. For example, most agents record only a partial negotiation history, while some acceptance strategies require the full history of the agent and/or its opponent. In such cases, the agent can be constrained to be incompatible with these acceptance strategies, or generalized to work with the full set of available acceptance strategies. We typically elected the most universal approach, even when this negatively influenced performance. We will demonstrate that while there is some performance loss when decoupling existing agents, it does not significantly impact the negotiation outcome.

The performance of the BOA agents was tested by letting them participate in a tournament with the same setup as ANAC 2011. The decoupled ANAC 2011 agents replaced the original agents, resulting in a tournament with eight participants. For the other BOA agents this was not possible, as their original counterparts did not participate in the ANAC 2011 competition. Therefore, for each of these agents we ran a modified tournament in which we added the original agent to the pool of ANAC 2011 agents, resulting in a tournament with nine participants. Next, we repeated this process for the BOA agents and evaluated the similarity of the results.

Table 3.2 ANAC 2011 reference results of the original agents using our hardware ($n = 10$)

Agent	Amsterdam	Camera	Car	Energy	Grocery	Company acquisition	Laptop	Nice or die	Mean utility
HardHeaded	0.891	**0.818**	**0.961**	0.664	0.725	0.747	0.683	**0.571**	**0.757**
Gahboninho	**0.912**	0.659	0.928	**0.681**	0.667	0.744	0.726	**0.571**	0.736
Agent K2	0.759	0.719	0.922	0.467	0.705	0.777	0.703	0.429	0.685
IAMhaggler 2011	0.769	0.724	0.873	0.522	0.725	**0.814**	**0.749**	0.300	0.685
BRAMAgent	0.793	0.737	0.815	0.420	0.724	0.744	0.661	**0.571**	0.683
The Negotiator	0.792	0.744	0.913	0.524	0.716	0.748	0.674	0.320	0.679
Nice Tit for Tat	0.733	0.765	0.796	0.508	0.759	0.767	0.660	0.420	0.676
ValueModelAgent	0.839	0.778	0.935	0.012	**0.767**	0.762	0.661	0.137	0.611

Best results are marked bold

For our experimental setup we used computers that were slower compared to the IRIDIS high-performance computing cluster that was used to run ANAC 2011. As we were therefore unable to reproduce exactly the same data as in Appendix D, we first recreated our own ANAC 2011 tournament data as depicted in Table 3.2, which is used as our baseline to benchmark the decoupled agents. The difference in performance caused small changes compared to the official ANAC 2011 ranking, causing *Agent K2* to move up from 5th to 3rd place.

Table 3.3 provides an overview of the results. We evaluated the performance in terms of the the difference in overall utility as well as the difference in time of agreement between the original and the BOA agents. The table does not list the agents that were not decoupled, and we also omitted *The Negotiator Reloaded* from the test set, as this agent was already submitted as a fully decoupled BOA agent.

From the results, we can conclude that the variation between the original and the BOA version is minimal; the majority of the standard deviations for both the difference in overall utility and time of agreement are close to zero. The largest difference between the original and decoupled agents with regard to the average time of agreement is 0.010 (*Agent Smith*); and for the average utility the largest difference is 0.015 (*BRAMAgent2*). Hence, in all cases the BOA agents and their original counterparts show comparable performance.

Table 3.3 Differences in overall utility and time of agreement between the original agents and their decoupled version

Agent	Diff. time agr.	SD time agr.	Diff. utility	SD utility
Agent K [13]	0.001	0.003	0.006	0.006
Agent Smith [14]	0.010	0.010	0.004	0.006
FSEGA [12]	0.001	0.004	0	0.003
IAMcrazyHaggler [8]	−0.004	0.012	0.003	0.013
IAMhaggler [8]	0.003	0.015	0.002	0.011
Nozomi	0.003	0.009	0.004	0.008
Yushu [18]	0.002	0.004	0.002	0.005
Agent K2 [19]	0.002	0.009	0.001	0.005
BRAMAgent [20]	0.004	0.011	0	0.006
Gahboninho [21]	0.001	0.008	0.006	0.005
HardHeaded [15]	−0.003	0.003	−0.009	0.004
IAMhaggler2011 [22]	−0.010	0.013	−0.002	0.003
Nice Tit for Tat [23]	0.006	0.010	−0.008	0.005
The Negotiator [24]	0	0.002	0	0.004
BRAMAgent2	0.002	0.011	−0.015	0.012
IAMhaggler2012	−0.005	0.006	−0.013	0.003
OMAC Agent [26]	0.003	0.003	0.012	0.015

Positive difference means the BOA agent performed better

3.4 Conclusion

This chapter introduces an architecture that distinguishes the bidding strategy, the opponent model, and the acceptance condition in negotiation agents, and recombines these components to systematically explore the space of automated negotiation strategies. The main idea behind the BOA architecture is that we can identify several components in a negotiating agent, all of which can be optimized individually. Our motivation in the end is to create a proficient negotiating agent by combining the best components.

We have shown that many of the existing negotiation strategies can be re-fitted into our architecture. We identified and classified the key components in them, and we have demonstrated that the original agents and their decoupled versions have identical behavior and similar performance.

With the BOA framework in place, the obvious direction to take is to analyze the BOA components in isolation, which we will do in the subsequent chapters. After identifying the best performing components, we answer in Chap. 10 whether combining effective components leads to better overall results, and whether an optimally performing agent can be created by taking the best of every component. We also answer the question which of the BOA components turns out to be most important with regard to the overall performance of an agent. Our architecture allows us to make these questions precise and provides a tool for answering them.

This chapter is based on the following publications: [31, 32]

Tim Baarslag, Koen V. Hindriks, Mark J.C. Hendrikx, Alex S.Y. Dirkzwager, and Catholijn M. Jonker. Decoupling negotiating agents to explore the space of negotiation strategies. In Ivan Marsa-Maestre, Miguel A. Lopez-Carmona, Takayuki Ito, Minjie Zhang, Quan Bai, and Katsuhide Fujita, editors, *Novel Insights in Agent-based Complex Automated Negotiation*, volume 535 of *Studies in Computational Intelligence*, pages 61–83. Springer, Japan, 2014

Tim Baarslag, Koen V. Hindriks, Mark J.C. Hendrikx, Alex S.Y. Dirkzwager, and Catholijn M. Jonker. Decoupling negotiating agents to explore the space of negotiation strategies. In *Proceedings of The Fifth International Workshop on Agent-based Complex Automated Negotiations (ACAN 2012)*, 2012

References

1. Lin R, Kraus S, Baarslag T, Tykhonov D, Hindriks KV, Jonker CM (2014) Genius: an integrated environment for supporting the design of generic automated negotiators. Comput Intell 30(1):48–70
2. Jonker CM, Treur J (2001) An agent architecture for multi-attribute negotiation. In: Proceedings of IJCAI'01, pp 1195–1201
3. Hindriks KV, Jonker CM, Tykhonov D (2007) Negotiation dynamics: analysis, concession tactics, and outcomes. In: Proceedings of the 2007 IEEE/WIC/ACM international conference on intelligent agent technology, IAT '07. IEEE Computer Society, Washington, DC, USA, pp 427–433

4. Lin R, Kraus S, Wilkenfeld J, Barry J (2008) Negotiating with bounded rational agents in environments with incomplete information using an automated agent. Artif Intell 172(6–7):823–851
5. Hindriks KV, Tykhonov D (2008) Opponent modelling in automated multi-issue negotiation using bayesian learning. In: Proceedings of the 7th international joint conference on autonomous agents and multiagent systems, AAMAS '08, vol 1. International foundation for autonomous agents and multiagent systems, Richland, SC, pp 331–338
6. Hindriks KV, Tykhonov D (2010) Towards a quality assessment method for learning preference profiles in negotiation. In: Ketter W, La Poutré JA, Sadeh N, Shehory O, Walsh W (eds) Agent-mediated electronic commerce and trading agent design and analysis, Lecture notes in business information processing, vol. 44. Springer, Berlin, pp 46–59
7. Williams CR, Robu V, Gerding EH, Jennings NR (2011) Using gaussian processes to optimise concession in complex negotiations against unknown opponents. In: Proceedings of the Twenty-second international joint conference on artificial intelligence, IJCAI'11, vol 1. AAAI Press, pp 432–438
8. Williams CR, Robu V, Gerding EH, Jennings NR (2012) Iamhaggler: a negotiation agent for complex environments. In: Ito T, Zhang M, Robu V, Fatima S, Matsuo T (eds) New trends in agent-based complex automated negotiations. Studies in computational intelligence. Springer, Berlin, pp 151–158
9. Frieder A, Miller G (2013) Value model agent: a novel preference profiler for negotiation with agents. In: Ito T, Zhang M, Robu V, Matsuo T (eds) Complex automated negotiations: theories. Models, and software competitions, vol 435. Studies in computational intelligence. Springer, Berlin, pp 199–203
10. Ilany L, Gal Y(K) (2014) The simple-meta agent. In: Marsa-Maestre I, Lopez-Carmona MA, Ito T, Zhang M, Bai Q, Fujita K (eds) Novel insights in agent-based complex automated negotiation. Studies in computational intelligence, vol 535. Springer, Japan, pp 197–200
11. Ilany L, Gal Y (2013) Algorithm selection in bilateral negotiation. In: Proceedings of the twenty-seventh AAAI conference on artificial intelligence (AAAI 2013)
12. Şerban LD, Silaghi GC, Litan CM (2012) AgentFSEGA—time constrained reasoning model for bilateral multi-issue negotiations. In: Ito T, Zhang M, Robu V, Fatima S, Matsuo T (eds) New trends in agent-based complex automated negotiations. Series of studies in computational intelligence. Springer, Berlin, pp 159–165
13. Kawaguchi S, Fujita K, Ito T (2012) AgentK: compromising strategy based on estimated maximum utility for automated negotiating agents. In: Ito T, Zhang M, Robu V, Fatima S, Matsuo T (eds) New trends in agent-based complex automated negotiations, Studies in computational intelligence, vol 383. Springer, Berlin, pp 137–144
14. van Galen Last N (2012) Agent Smith: opponent model estimation in bilateral multi-issue negotiation. In: Ito T, Zhang M, Robu V, Fatima S, Matsuo T (eds) New trends in agent-based complex automated negotiations. Studies in computational intelligence. Springer, Berlin, pp 167–174
15. van Krimpen T, Looije D, Hajizadeh S (2013) Hardheaded. In: Ito T, Zhang M, Robu V, Matsuo T (eds) Complex automated negotiations: theories, models, and software competitions. Studies in computational intelligence, vol 435. Springer, Berlin, pp 223–227
16. Hao J, Leung H-F (2012) ABiNeS: an adaptive bilateral negotiating strategy over multiple items. In: Proceedings of the the 2012 IEEE/WIC/ACM international joint conferences on web intelligence and intelligent agent technology, WI-IAT '12, vol 2. IEEE Computer Society, Washington, DC, USA, pp 95–102
17. Hao J, Leung H (2014) CUHK agent: an adaptive negotiation strategy for bilateral negotiations over multiple items. In: Marsa-Maestre I, Lopez-Carmona MA, Ito T, Zhang M, Bai Q, Fujita K (eds) Novel insights in agent-based complex automated negotiation, Studies in computational intelligence, vol 535. Springer, Japan, pp 171–179
18. An B, Lesser VR (2012) Yushu: a heuristic-based agent for automated negotiating competition. In: Ito T, Zhang M, Robu V, Fatima S, Matsuo T (eds) New trends in agent-based complex automated negotiations, Studies in Computational Intelligence, vol 383. Springer, Berlin, pp 145–149

19. Kawaguchi S, Fujita K, Ito T (2013) AgentK2: compromising strategy based on estimated maximum utility for automated negotiating agents. In: Ito T, Zhang M, Robu V, Matsuo T (eds) Complex automated negotiations: theories, models, and software competitions, Studies in computational intelligence, vol 435. Springer, Berlin, pp 235–241

20. Fishel R, Bercovitch M, Gal Y(K) (2013) Bram agent. In: Ito T, Zhang M, Robu V, Matsuo T (eds) Complex automated negotiations: theories, models, and software competitions. Studies in computational intelligence, vol 435. Springer, Berlin, pp 213–216

21. Adar MB, Sofy N, Elimelech A (2013) Gahboninho: strategy for balancing pressure and compromise in automated negotiation. In: Ito T, Zhang M, Robu V, Matsuo T (eds) Complex automated negotiations: theories, models, and software competitions. Studies in computational intelligence, vol 435. Springer, Berlin, pp 205–208

22. Williams CR, Robu V, Gerding EH, Jennings NR (2013) Iamhaggler 2011: a gaussian process regression based negotiation agent. In: Ito T, Zhang M, Robu V, Matsuo T (eds) Complex automated negotiations: theories, models, and software competitions. Studies in computational intelligence, vol 435. Springer, Berlin, pp 209–212

23. Baarslag T, Hindriks KV, Jonker CM (2013) A tit for tat negotiation strategy for real-time bilateral negotiations. In: Ito T, Zhang M, Robu V, Matsuo T (eds) Complex automated negotiations: theories, models, and software competitions. Studies in computational intelligence, vol 435. Springer, Berlin, pp 229–233

24. Dirkzwager ASY, Hendrikx MJC, Ruiter JR (2013) The negotiator: a dynamic strategy for bilateral negotiations with time-based discounts. In: Ito T, Zhang M, Robu V, Matsuo T (eds) Complex automated negotiations: theories, models, and software competitions. Studies in computational intelligence, vol 435. Springer, Berlin, pp 217–221

25. Morii S, Ito T (2014) AgentMR: concession strategy based on heuristic for automated negotiating agents. In: Marsa-Maestre I, Lopez-Carmona MA, Ito T, Zhang M, Bai Q, Fujita K (eds) Novel insights in agent-based complex automated negotiation. Studies in computational intelligence, vol 535. Springer, Japan, pp 181–186

26. Chen S, Weiss G (2012) An efficient and adaptive approach to negotiation in complex environments. In: De Raedt L, Bessiere C, Dubois D, Doherty P, Frasconi P, Heintz F, Lucas P (eds) ECAI. Frontiers in artificial intelligence and applications, vol 242. IOS Press, Amsterdam, pp 228–233

27. Kawaguchi S, Fujita K, Ito T (2011) Compromising strategy based on estimated maximum utility for automated negotiation agents competition (ANAC-10). In: Mehrotra KG, Mohan CK, Oh JC, Varshney PK, Ali M (eds) Modern approaches in applied intelligence. Lecture notes in computer science, vol 6704. Springer, Berlin, pp 501–510

28. Zeng D, Sycara KP (1998) Bayesian learning in negotiation. Int J Hum Comput Stud 48(1): 125–141

29. Faratin P, Sierra C, Jennings NR (1998) Negotiation decision functions for autonomous agents. Robot Auton Syst 24(3–4):159–182

30. Fatima SS, Wooldridge MJ, Jennings NR (2002) Multi-issue negotiation under time constraints. In: AAMAS '02: proceedings of the first international joint conference on autonomous agents and multiagent systems. ACM, New York, pp 143–150

31. Baarslag T, Hindriks KV, Hendrikx MJC, Dirkzwager ASY, Jonker CM (2014) Decoupling negotiating agents to explore the space of negotiation strategies. In: Marsa-Maestre I, Lopez-Carmona MA, Ito T, Zhang M, Bai Q, Fujita K (eds) Novel insights in agent-based complex automated negotiation. Studies in computational intelligence, vol 535. Springer, Japan, pp 61–83

32. Baarslag T, Hindriks KV, Hendrikx MJC, Dirkzwager ASY, Jonker CM (2012) Decoupling negotiating agents to explore the space of negotiation strategies. In: Proceedings of the fifth international workshop on agent-based complex automated negotiations (ACAN 2012)

Chapter 4
Effective Acceptance Conditions

Abstract An essential part of our framework outlined in Chap. 3 is the acceptance strategy of an agent. In every negotiation with a deadline, one of the negotiating parties must accept an offer to avoid a break off. As a break off is usually an undesirable outcome for both parties, it is important that a negotiator employs a proficient mechanism to decide under which conditions to accept. When designing such conditions, one is faced with *the acceptance dilemma*: accepting the current offer may be suboptimal, as better offers may still be presented before time runs out. On the other hand, accepting too late may prevent an agreement from being reached, resulting in a break off with no gain for either party. Motivated by the challenges of bilateral negotiations between automated agents and by the results and insights of the automated negotiating agents competition of 2010, we *classify* and *compare* state-of-the-art generic acceptance conditions in this chapter. We perform extensive experiments to compare the performance of various acceptance conditions in combination with a broad range of bidding strategies and negotiation scenarios. Furthermore, we propose new acceptance conditions and we demonstrate that they outperform the other conditions. We also provide insight into why some conditions work better than others and investigate correlations between the properties of the negotiation scenario and the efficacy of acceptance conditions.

4.1 Introduction

In 2010, seven new negotiation strategies were created to participate in the first automated negotiating agents competition (ANAC 2010, see Appendix C) in conjunction with the Ninth International Conference on Autonomous Agents and Multiagent Systems (AAMAS-10). During post tournament analysis of the results, it became apparent that different agent implementations use various conditions to decide when to accept an offer. It is important for every negotiator to employ such a mechanism to decide under which conditions to accept, because in every negotiation with a deadline, one of the negotiating parties has to accept in order to avoid a break off. However, designing a proper acceptance condition is a difficult task: accepting too late may result in the break off of a negotiation, while accepting too early may result in suboptimal agreements.

© Springer International Publishing Switzerland 2016
T. Baarslag, *Exploring the Strategy Space of Negotiating Agents*,
Springer Theses, DOI 10.1007/978-3-319-28243-5_4

Table 4.1 An overview of the rank of every agent in ANAC 2010 and the type of acceptance conditions that they employ

Rank	Agent	Acceptance condition
1	*Agent K*	Time and utility based
2	*Yushu*	Time and utility based
3	*Nozomi*	Time and utility based
4	*IAMhaggler*	Utility based only
5	*FSEGA*	Utility based only
6	*IAMcrazyHaggler*	Utility based only
7	*Agent Smith*	Time and utility based

Agents using time and utility based acceptance conditions were ranked at the top, except for *Agent Smith*, which had a faulty acceptance mechanism

The importance of choosing an appropriate acceptance condition is confirmed by the results of ANAC 2010 (see Table 4.1). Agents with simple acceptance conditions were ranked at the bottom, while the more sophisticated time- and utility-based conditions obtained a higher score. For instance, the low ranking of *Agent Smith* was due to a mistake in the implementation of the acceptance condition [1].

Despite its importance, the theory and practice of acceptance conditions has not yet received much attention. The goal of this chapter is to classify current approaches and to compare acceptance conditions in an experimental setting. Thus in this chapter we will concentrate on the final part of the negotiation process: the acceptation of an offer. We focus on *decoupled* acceptance conditions: i.e., generic acceptance conditions that can be used in conjunction with an arbitrary bidding strategy and hence, fit into the BOA architecture described in Chap. 3. The reason for this is straightforward: we want to be able to re-incorporate the acceptance conditions that have been found most effective into new agent designs; therefore, the acceptance conditions under investigation should not be coupled with a specific agent implementation.

The contribution of this chapter is fourfold:

1. We give an overview and provide a categorization of current decoupled acceptance conditions.
2. We introduce a formal negotiation model that supports the use of arbitrary acceptance conditions.
3. We compare a large selection of current generic acceptance conditions and evaluate them in an experimental setting.
4. We propose new acceptance conditions and test them against established acceptance conditions, using varying types of bidding techniques.

The remainder of this chapter is organized as follows. Section 4.2 defines a formal model of accepting in negotiation and provides an overview of current acceptance conditions. In Sect. 4.3, we also consider combinations of acceptance conditions. Section 4.4 discusses our experimental setup and results, which demonstrate that some combinations outperform traditional acceptance conditions. Finally, Sect. 4.5 outlines the conclusions of this chapter.

4.2 Acceptance Conditions in Negotiation

We focus on acceptance conditions that are decoupled: i.e. generic acceptance conditions that are not tied to a specific agent implementation and hence can be used in conjunction with an arbitrary bidding strategy. We first define a general negotiation model that fits current decoupled acceptance conditions. We have surveyed existing negotiation agents to examine the acceptance conditions that they employ. We then categorize them according to the input that they use in their decision making process.

4.2.1 A Formal Model of Accepting

We briefly review our definitions of Sect. 2.2.2. The interaction between the agents is regulated by the alternating-offers protocol supplemented with a real time line $\mathcal{T}$, represented here by $\mathcal{T} = [0, 1]$, so that the deadline occurs at $t = 1$. This is the same setup as [2], with the exception that issues are not necessarily real-valued and both agents have the same deadline.

We represent by $x_{A \to B}^t$ the negotiation outcome proposed by agent A to agent B at time t. A *negotiation thread* (cf. [3, 4]) between two agents A and B at time $t \in \mathcal{T}$ is defined as a finite sequence

$$H_{A \leftrightarrow B}^t := \left(x_{p_1 \to p_2}^{t_1}, x_{p_2 \to p_3}^{t_2}, x_{p_3 \to p_4}^{t_3}, \ldots, x_{p_n \to p_{n+1}}^{t_n} \right), \tag{4.1}$$

which satisfies the following constraints:

1. The offers are ordered over time $\mathcal{T}: t_k \leq t_l$ for $k \leq l$.
2. The offers are alternating between the agents: $p_k = p_{k+2} \in \{A, B\}$ for all k.
3. All t_i represent instances of time $\mathcal{T}$, with $t_n \leq t$.
4. The agents exchange complete offers: $x_{p_k \to p_{k+1}}^{t_k} \in \Omega$ for $k \in \{1, \ldots, n\}$.

The last element of $H_{A \leftrightarrow B}^t$ may also be equal to *Accept* or *End*. We will say a negotiation thread is *active* if this is not the case.

We now formally define how an agent reaches the decision to accept. When agent A receives an offer $x_{B \to A}^t$ from agent B sent at time t, it has to decide at a later time $t' > t$ whether to accept the offer, or to send a counter-offer $x_{A \to B}^{t'}$. Given a negotiation thread $H_{A \leftrightarrow B}^t$ between agents A and B, we can formally express the action performed by A with an *action function* X_A:

$$X_A(t', x_{B \to A}^t) = \begin{cases} End & \text{if } t' \geq 1 \\ Accept & \text{if } \mathbf{AC}_A(t', x_{A \to B}^{t'}, H_{A \leftrightarrow B}^t) \\ Offer \ x_{A \to B}^{t'} & \text{otherwise} \end{cases} \tag{4.2}$$

Note that we extend the setting of [2, 4] by introducing the *acceptance condition* $\mathbf{AC}_A$ of an agent A. When used in this way, the model enables us to study arbitrary decoupled acceptance conditions. The acceptance condition $\mathbf{AC}_A$ takes as input

$$\mathcal{I} = (t', x^{t'}_{A \to B}, H^t_{A \leftrightarrow B}), \tag{4.3}$$

the tuple containing the current time t', the offer $x^{t'}_{A \to B}$ that the agent considers as a bid (in line with the bidding strategy the agent uses), and the active negotiation thread $H^t_{B \leftrightarrow A}$.

The resulting action given by the function $X_A(t', x^t_{B \to A})$ is used to extend the current negotiation thread between the two agents. If the agent does not accept the current offer, and the deadline has not been reached, it will prepare a counter-offer $x^{t'}_{A \to B}$ by using a bidding strategy or *tactic* to generate new values for the negotiable issues. As explained in Sect. 2.3.3, tactics can take many forms, e.g. time-dependent, resource dependent, imitative, and so on [4]. In our setup we will consider the tactics as given and try to optimize the accompanying acceptance conditions.

4.2.2 Acceptance Conditions

Let an active negotiation thread

$$H^t_{A \leftrightarrow B} = \left(x^{t_1}_{p_1 \to p_2}, x^{t_2}_{p_2 \to p_3}, \ldots, x^{t_{n-1}}_{A \to B}, x^{t_n}_{B \to A} \right),$$

be given at time $t' > t = t_n$, so that it is agent A's turn to perform an action.

The action function X_A of an agent A uses an acceptance condition $\mathbf{AC}_A(\mathcal{I})$ to decide whether to accept, as defined by Eq. (4.2). In practice, most agents do not use the full negotiation thread to decide whether it is time to accept. For instance many agent implementations, such as [2, 4, 5], use the following implementation of $\mathbf{AC}_A(\mathcal{I})$:

$$\mathbf{AC}_A(t', x^{t'}_{A \to B}, H^t_{A \leftrightarrow B}) \iff u_A(x^t_{B \to A}) \geq u_A(x^{t'}_{A \to B}).$$

That is, A will accept when the utility u_A of the opponent's last offer at time t is greater than the value of the offer agent A is ready to send out at time t'. The acceptance condition above depends on the agent's upcoming offer $x^{t'}_{A \to B}$. For $\alpha, \beta \in \mathbb{R}$ this may be generalized as follows:

$$\mathbf{AC}^{\mathcal{I}}_{\text{next}}(\alpha, \beta) \overset{\text{def}}{\iff} \alpha \cdot u_A(x^t_{B \to A}) + \beta \geq u_A(x^{t'}_{A \to B}). \tag{4.4}$$

We can view α as the scale factor by which we multiply the opponent's bid, while β specifies the minimal 'utility gap' [6] that is sufficient to accept.

Analogously, we have acceptance conditions [6–9] that rely on the agent's *previous* offer $x^{t_{n-1}}_{A \to B}$:

$$\mathbf{AC}^{\mathcal{I}}_{\text{prev}}(\alpha, \beta) \overset{\text{def}}{\iff} \alpha \cdot u_A(x^t_{B \to A}) + \beta \geq u_A(x^{t_{n-1}}_{A \to B}). \tag{4.5}$$

Note that this acceptance condition does not take into account the time that is left in the negotiation, nor any offers made previous to time t. However, it is important to bear in mind that the behavior of the acceptance condition may still be influenced implicitly by these factors, because of the possibility that the bidding strategy takes such factors into account.

Other acceptance conditions may rely on other measures, such as the remaining negotiation time or a utility threshold. For example, there is a very simple acceptance criterion [1, 8, 9] that only compares the opponent's previous offer with a threshold α:

$$\mathbf{AC}_{\mathrm{const}}^{\mathcal{I}}(\alpha) \stackrel{\mathrm{def}}{\Longleftrightarrow} u_A(x_{B \to A}^t) \geq \alpha. \tag{4.6}$$

Last but not least, instead of considering utility, agents (such as [1]) may employ a time-based condition to accept after a certain amount of time $T \in \mathcal{T}$ has passed:

$$\mathbf{AC}_{\mathrm{time}}^{\mathcal{I}}(T) \stackrel{\mathrm{def}}{\Longleftrightarrow} t' \geq T. \tag{4.7}$$

We will omit the superscript $\mathcal{I}$ in Eqs. (4.4)–(4.7) when it is clear from the context. We will use these general acceptance conditions to classify existing acceptance mechanisms in the next section.

4.2.3   Existing Acceptance Conditions

We give a short overview of decoupled acceptance conditions used in literature and current agent implementations. We are primarily interested in acceptance conditions that are not specifically designed for a single agent. We do not claim the list below is complete; however it serves as a good starting point to categorize current decoupled acceptance conditions. We surveyed the entire pool of agents of ANAC 2010, including *Agent K, Nozomi* [10], *Yushu* [11], *IAM(crazy)Haggler* [9], *FSEGA* [8], and *Agent Smith* [1]. We also examined well-known agents from literature, such as the *Trade-off agent* [12], the *Bayesian learning agent* [7], *ABMP* [6], equilibrium strategies of [5], and time-dependent negotiation strategies as defined in Sect. 2.3.3, i.e., the *Boulware* and *Conceder* tactics.

Listed in Table 4.2 is a selection of generic acceptance conditions found.

Some agents also use logical combinations of different acceptance conditions at the same time. This explains why some agents are listed multiple times in the table. For example, both *IAMHaggler* and *IAMcrazyHaggler* [9] accept precisely when

$$\mathbf{AC}_{\mathrm{const}}(0.88) \vee \mathbf{AC}_{\mathrm{next}}(1.02, 0) \vee \mathbf{AC}_{\mathrm{prev}}(1.02, 0).$$

We will not focus on the many possible combinations of all acceptance conditions that may thus be obtained; we will study the basic acceptance conditions in isolation with varying parameters. However in addition to this, we study a small selection of combinations in Sect. 4.3. We leave further combinations for future research.

Table 4.2 A selection of existing decoupled acceptance conditions found in literature and current agent implementations

Acceptance condition	α	β	T	Agent
$\mathbf{AC}_{\text{prev}}(\alpha, \beta)$	1.03	0	–	FSEGA, Bayesian Agent
	1	0	–	Agent Smith
	1.02	0	–	IAM(crazy)Haggler
	1	0.02	–	ABMP
$\mathbf{AC}_{\text{next}}(\alpha, \beta)$	1	0	–	FSEGA, Boulware, Conceder, Trade-off, Equilibrium strategies
	1.02	0	–	IAM(crazy)Haggler
	1.03	0	–	Bayesian Agent
$\mathbf{AC}_{\text{const}}(\alpha)$	1	–	–	FSEGA
	0.9	–	–	Agent Smith
	0.88	–	–	IAM(crazy)Haggler
$\mathbf{AC}_{\text{time}}(T)$	–	–	0.92	Agent Smith

As can be seen from Table 4.2, the most commonly used acceptance condition in our sample is $\mathbf{AC}_{\text{next}} = \mathbf{AC}_{\text{next}}(1, 0)$, which is the familiar condition of accepting when the opponent's last offer is better than the planned offer of the agent. The function $\beta \mapsto \mathbf{AC}_{\text{prev}}(1, \beta)$ can be viewed as an acceptance condition that accepts when the *utility gap* [6] between the parties is smaller than β. We denote this condition by $\mathbf{AC}_{\text{gap}}(\beta)$.

4.3 Combined Acceptance Conditions

We define three acceptance conditions that are designed to perform well in conjunction with an arbitrary bidding strategy. This will incorporate all ideas behind the traditional acceptance conditions we have described so far. We will show in Sect. 4.4 that they work better than the majority of simple generic conditions listed in Table 4.2.

From a negotiation point of view, it makes sense to alter the behavior of an acceptance condition when time is running short. For example, many ANAC agents such as *Yushu*, *Nozomi* and *FSEGA* [8, 10, 11] split the negotiation into different intervals of time and apply different sub-strategies to each interval.

The basic idea behind combined acceptance conditions $\mathbf{AC}_{\text{combi}}$ is similar. In case the bidding strategy plans to propose a deal that is worse than the opponent's offer, we have reached a consensus with our opponent and we accept the offer. However, if there still exists a gap between our offer and time is short, the acceptance condition should wait for an offer that is not expected to improve in the remaining time. Thus $\textbf{AC}_{\text{combi}}$ is designed to be a proper extension of $\textbf{AC}_{\text{next}}$, with adaptive behavior based on recent bidding behavior near the deadline.

To define $\textbf{AC}_{\text{combi}}$, suppose an active negotiation thread

$$H^t_{A \leftrightarrow B} = \left(x^{t_1}_{p_1 \to p_2}, x^{t_2}_{p_2 \to p_3}, \ldots, x^{t_{n-1}}_{A \to B}, x^{t_n}_{B \to A} \right),$$

is given at time $t' > t = t_n > \frac{1}{2}$ near the deadline, when it is agent A's turn. Note that there is $r = 1 - t'$ time remaining in the negotiation, which we will call the *remaining time window*. A good sample of what might be expected in the remaining time window consists of the bids that were exchanged during the previous time window $W = [t' - r, t'] \subseteq \mathcal{T}$ of the same size.

Let

$$H^W_{B \to A} = \left\{ x^s_{B \to A} \in H^t_{A \leftrightarrow B} \mid s \in W \right\}$$

denote all bids offered by B to A in time window W. We can now formulate the average and maximum utility that was offered during the previous time window in the negotiation thread $H = H^W_{B \to A}$:

$$\text{MAX}^W = \max_{x \in H} u_A(x),$$

and

$$\text{AVG}^W = \frac{1}{|H|} \sum_{x \in H} u_A(x).$$

We let $\textbf{AC}_{\text{combi}}(T, \alpha)$ accept at time t' exactly when the following holds: $\textbf{AC}_{\text{next}}$ indicates that we have to accept, *or* we have almost reached the deadline ($t' \geq T$) and the current offer suffices (i.e. better than α) given the remaining time:

$$\textbf{AC}_{\text{combi}}(T, \alpha)$$
$$\overset{\text{def}}{\Longleftrightarrow} \tag{4.8}$$
$$\textbf{AC}_{\text{next}} \vee \textbf{AC}_{\text{time}}(T) \wedge \left(u_A(x^t_{B \to A}) \geq \alpha \right).$$

Note that Eq. (4.8) defines $\textbf{AC}_{\text{combi}}(T, \alpha)$ in such a way that it splits the negotiation time into two phases: $[0, T)$ and $[T, 1]$, with different behavior in both cases.

We will consider three different combined acceptance conditions:

1. $\textbf{AC}_{\text{combi}}(T, \text{MAX}^W)$: the current offer is good enough when it is better than all offers seen in the previous time window W,
2. $\textbf{AC}_{\text{combi}}(T, \text{AVG}^W)$: the offer is better than the average utility of offers during the previous time window W,
3. $\textbf{AC}_{\text{combi}}(T, \text{MAX}^T)$: the offer should be better than any bid seen before.

4.4 Experiments

In order to experimentally test the efficacy of an acceptance condition, we considered
a negotiation setup with the following characteristics. We equipped a set of agents
(as defined later) with an acceptance condition, and measured the result against
other agents in the following way. Suppose agent A is equipped with acceptance
condition AC_A and negotiates with agent B. The two parties may reach a certain
outcome $\omega \in \Omega$, for which A receives the associated utility $u_A(\omega)$. The score for A
is averaged over all trials on various domains (see Sect. 4.4.1), alternating between
the two preference profiles defined on that domain. E.g., on the negotiation scenario
between *England* and *Zimbabwe*, A will play both as *England* and as *Zimbabwe*
against all others. This average utility score is then an indication of the efficacy
of AC_A.

For our experimental setup we employed GENIUS as described in Appendix A. It
supports the alternating offer protocol with a real-time deadline as outlined in our
negotiation model. The default negotiation time in GENIUS and in the setup of ANAC
is 3 min per negotiation session; therefore we use the same value in our experiments.

4.4.1 Detailed Experimental Setup

4.4.1.1 Agents

We use the negotiation strategies that were submitted to The Automated Negotiating
Agents Competition (ANAC 2010, see Appendix C). The seven agents that participated in ANAC 2010 were implemented by various international research groups
of negotiation experts. Firstly, we removed the built-in acceptance mechanism from
this representative group of agents; this left us with its pure bidding strategies. In
terms of our BOA architecture described in Chap. 3, we replaced the acceptance
strategy of the agents, but we left the bidding strategy and opponent modeling component intact. As outlined in our negotiation model, this procedure allowed us to test
arbitrary acceptance conditions in tandem with any ANAC tactic.

We aimed to tune our acceptance conditions to the top performing ANAC 2010
agents. Therefore we selected the top 3 of ANAC agents that were submitted by
different research groups, namely *Agent K*, *Yushu* and *IAMhaggler* (we omitted
Nozomi as the designing group also implemented *Agent K*, cf. Table 4.1). For the
set of opponents, we selected all agents from ANAC 2010, for the acceptance conditions should be tested against a wide array of strategies. The opponents also had
their built-in acceptance conditions removed (and hence were not able to accept), so
that differences in results would depend entirely on the acceptance condition under
consideration. To test the efficacy of an acceptance condition, we equipped the top 3
bidding strategies with this condition and compared the average utility obtained by
the three agents when negotiating against their opponents.

Table 4.3 The eight preference profiles from ANAC 2010 and ANAC 2011, as used in the experiments

	Laptop	Itex–Cyp	Eng–Zim	Grocery
Size	27	180	576	1600
Opposition	Weak	Strong	Medium	Medium
Mean utility	0.67	0.48	0.58	0.44
Nash point	(1.00, 0.82)	(0.72, 0.67)	(0.91, 0.73)	(0.84, 0.90)
K-S point	(0.87, 0.87)	(0.72, 0.67)	(0.82, 0.79)	(0.84, 0.90)

The rows indicate respectively: the size of the outcome space, the level of opposition, the arithmetic mean utility that can be obtained in the scenario, and the location of the Nash point and Kalai Smorodinsky point

4.4.1.2 Domains

The specifics of a negotiation domain can be of great influence on the negotiation outcome [13]. Acceptance conditions have to be assessed on negotiation domains of different size and complexity. Negotiation results also depend on the *opposition* of the parties' preferences (see Sect. 2.2.4). Strong opposition is typical of competitive domains, when a gain for one party can be achieved only at a loss for the other party. Conversely, weak opposition means that both parties achieve either losses or gains simultaneously.

With this in mind, we aimed for a good spread of negotiation characteristics by selecting four different negotiation scenarios with two preference profiles each (see Table 4.3). We picked two domains from the three that were used in ANAC 2010 (cf. Appendix C). We also selected two negotiation scenarios from the ANAC 2011 competition (cf. Appendix D) to include both a smaller and a larger domain to our experimental setup.

We omitted the largest domains that featured in ANAC 2010 and 2011, as some ANAC 2010 agents did not scale well and had too many difficulties to make these scenarios reliable for testing. Additionally, in contrast to the 2010 competition, ANAC 2011 introduced discount factors for some of the scenarios. We removed these discount factors to ensure compatibility with the ANAC 2010 agents.

We shortly describe our negotiation scenarios here; for more detailed information, we refer to Appendix C and D.

Our smallest scenario is called *Laptop*. In this scenario, a seller and a buyer, negotiate over the specifications of a laptop. There are three issues: the laptop brand, the size of the hard disk, and the size of the external monitor. Each issue has only three options, making 27 possible outcomes. If the two parties are able to find the outcomes that are mutually beneficial to both, then they are happy to do business together with high utility scores on both sides. This can be confirmed in Table 4.3: the scenario has the highest arithmetic mean utility, and the most favorable Nash and Kalai-Smorodinsky point.

Our second scenario is taken from [14], which describes a buyer–seller business negotiation. There are four issues that both sides have to discuss: the price of the components, delivery times, payment arrangements and terms for the return of possibly defective parts. The opposition between the parties is strong in this domain, as the manufacturer and consumer have naturally opposing requirements. Even the Nash point utilities are quite low for both parties. Altogether, there are 180 potential offers that contain all combinations of values for the four issues.

Third, the domain taken from [15, 16] involves a case where *England* and *Zimbabwe* negotiate an agreement on tobacco control. The leaders of both countries must reach an agreement on five issues. *England* and *Zimbabwe* have contradictory preferences for the first two issues, but the other issues have options that are jointly preferred by both sides. The domain has a total of 576 possible agreements.

Our final negotiation case concerns the *Grocery* scenario, which models a shopping negotiation in a local supermarket. The negotiation is between two persons having different tastes, who wish to buy groceries together. The discussion is about five product categories with four to five possible options, resulting in a scenario with 1600 possible outcomes. The preferences are modeled in such a way that a good outcome is achievable for both, so the Nash and Kalai-Smorodinsky point utilities are high for both parties; however, the outcome space is scattered (resulting in a relatively low mean utility), so agents must explore it considerably to find the jointly profitable ones.

To compensate for any utility differences in the preference profiles, the agents play both sides of every scenario.

4.4.1.3 Acceptance Conditions

For each acceptance condition we tested all $3 \times 7 = 21$ pairings of agents, playing with each of the 8 different preference profiles. We ran every experiment a total of $N = 15$ times, so that altogether each acceptance condition was tested $21 \times 8 \times 15 = 2520$ times in total. This resulted in running as many negotiations, and as every negotiation lasts 3 min, the experiments took 126 hours of cpu time. We selected a wide range of 102 acceptance conditions for experimental testing, as shown in Table 4.4. The different values of parameters will be discussed in the section below.

Additionally, we ran five more experiments with agents having their original, built-in acceptance mechanism in place. That is, we also tested the original agents' *coupled* acceptance mechanism for comparison purposes. As we cannot for example, equip *Agent K* with the coupled acceptance condition of *Yushu*, we tested the built-in mechanism by having each agent employ its own mechanism.

4.4.2 Hypotheses and Experimental Results

The experiments considered here are designed to discuss the main properties and drawbacks of the acceptance conditions listed above. We formulate several hypotheses with respect to the acceptance conditions we have discussed.

Table 4.4 The selected ranges and increments for the parameters of different acceptance conditions in the experimental setup

Acceptance condition	Ranges	Increments
$\mathbf{AC}_{\text{prev}}(\alpha, \beta)$	$\alpha \in [1, 1.05), \beta \in [0, 0.1)$	for α: 0.01, for β: 0.02
$\mathbf{AC}_{\text{next}}(\alpha, \beta)$	$\alpha \in [1, 1.05), \beta \in [0, 0.1)$	for α: 0.01, for β: 0.02
$\mathbf{AC}_{\text{const}}(\alpha)$	$\alpha \in [0, 1)$	0.05
$\mathbf{AC}_{\text{time}}(T)$	$T \in [0, 1)$	0.05
$\mathbf{AC}_{\text{combi}}\left(T, \text{MAX}^W\right)$	$T \in [0.95, 1)$	0.01
$\mathbf{AC}_{\text{combi}}\left(T, \text{AVG}^W\right)$	$T \in [0.95, 1)$	0.01
$\mathbf{AC}_{\text{combi}}\left(T, \text{MAX}^T\right)$	$T = 0.99$	–

To evaluate the hypotheses below, we have carried out a large number of experiments. A small selection of the results is summarized in Table 4.5. The table shows the average utility obtained by the agents, and the standard deviation (of the $N = 15$ experiments), when equipped with several acceptance conditions. The "average utility of agreements" column represents the average utility obtained by the agent given the fact that they have reached an agreement. When they do not reach an agreement (due to reaching the deadline), they get zero utility. Thus, as a general observation, the following holds:

(The acceptance dilemma)
Total average utility = Agreement percentage × Average utility of agreements.

This formula captures the essence of the acceptance dilemma: accepting bad to mediocre offers yields more agreements of relatively low utility; while accepting only the best offers produces less agreements, but of higher utility.

Our first hypothesis is about the simplest condition, $\mathbf{AC}_{\text{const}}(\alpha)$, and reads as follows:

Hypothesis 4.1 There is no single choice for α that makes $\mathbf{AC}_{\text{const}}(\alpha)$ an effective acceptance condition; this is mainly because the optimal choice of α is very domain-dependent.

First, consider $\mathbf{AC}_{\text{const}}(0.9)$ and $\mathbf{AC}_{\text{const}}(0.8)$ by consulting Table 4.5. When they reach an agreement, they receive a very high utility (at least 0.9 or 0.8 respectively), but this happens so infrequently (resp. 60 and 36 % of all negotiations), that they are ranked at the bottom when we consider total average utility. On the other hand, choosing a low value for α, such as using $\mathbf{AC}_{\text{const}}(0.2)$, will always result in an immediate agreement, but with one of the lowest possible scores of 0.492.

Table 4.5 A small selection of the various acceptance conditions that were tested, together with average utility obtained and standard deviation

Acceptance condition	α	β	T	Utility	SD	Agt. (%)	Agt. utility
$AC_{prev}(\alpha, \beta)$	1	0	–	0.680	0.0084	80	0.851
	1	0.04	–	0.711	0.0094	84	0.842
	1	0.08	–	0.722	0.0076	87	0.827
	1.02	0.04	–	**0.723**	0.0085	86	0.837
$AC_{next}(\alpha, \beta)$	1	0	–	0.683	0.0112	81	0.843
	1	0.04	–	0.727	0.0067	87	0.833
	1	0.08	–	0.731	0.0057	89	0.819
	1.02	0.04	–	**0.737**	0.0060	89	0.830
$AC_{const}(\alpha)$	0.20	–	–	0.492	0.0025	100	0.492
	0.55	–	–	**0.619**	0.0027	92	0.671
	0.80	–	–	0.501	0.0078	60	0.842
	0.90	–	–	0.343	0.0080	36	0.952
Built-in mechanism	–	–	–	**0.737**	0.0057	89	0.774
$AC_{time}(T)$	–	–	0.10	0.533	0.0035	100	0.533
	–	–	0.40	0.548	0.0064	100	0.548
	–	–	0.70	0.602	0.0062	100	0.602
	–	–	0.95	**0.648**	0.0063	100	0.648
$AC_{combi}(T, MAX^W)$	–	–	0.97	0.756	0.0019	100	0.756
	–	–	0.98	**0.762**	0.0031	100	0.764
	–	–	0.99	0.761	0.0046	98	0.776
$AC_{combi}(T, AVG^W)$	–	–	0.97	0.739	0.0050	100	0.739
	–	–	0.98	0.754	0.0037	100	0.757
	–	–	0.99	**0.759**	0.0056	98	0.774
$AC_{combi}(T, MAX^T)$	–	–	0.99	**0.737**	0.0083	93	0.796

The utility of the best scoring AC of each category is represented in **bold**. The two right-hand side columns show agreement percentages and the utility obtained when an agreement is reached

The best possible choice for α should therefore be somewhere in the middle between zero and one, and is found to be 0.55 (see Fig. 4.1), yielding a payoff of 0.619. Firstly, this is still a suboptimal outcome compared to other AC's, such as the AC_{next} and AC_{combi} variants.

Moreover, it is worth noting that this optimal value may be best on average, but in this case, averaging over all scenarios also hides a lot of information. When we break down our analysis and look at the four domains separately (see the four figures of Fig. 4.2), we see that the optimal range of α differs greatly per domain. For example, on *Itex versus Cypress*, the optimal choice for α is around 0.6, while on *Grocery*, the best performing value is in the range of [0.7, 0.8]. On the *Laptop* domain, any choice for $\alpha \in [0, 0.8]$ is the best $AC_{const}(\alpha)$ can do in this scenario, and will cause the agent to instantly accept most offers.

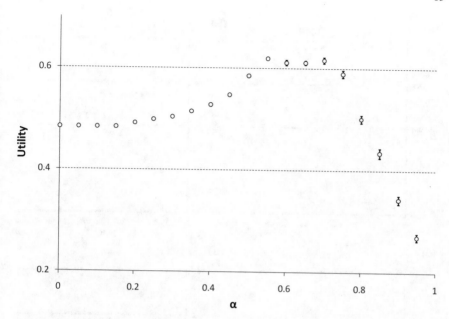

Fig. 4.1 The average utility obtained by agents using $\mathbf{AC}_{\text{const}}(\alpha)$. The *vertical error bars* indicate one standard deviation to the mean

We conclude that our hypothesis is confirmed: in isolation, $\mathbf{AC}_{\text{const}}(\alpha)$ is not very advantageous to use. The main reason is that the choice of the constant α is highly domain-dependent. A very cooperative scenario may have multiple win–win outcomes with utilities above α. $\mathbf{AC}_{\text{const}}(\alpha)$ would then accept an offer which is *relatively* bad, i.e. it could have done much better. On the other hand, in highly competitive domains, it may simply 'ask for too much' and may rarely obtain an agreement. Its value lies mostly in using it in combination with other acceptance conditions such as $\mathbf{AC}_{\text{next}}$. It can then benefit the agent by accepting an unexpectedly good offer or a mistake by the opponent.

As we discussed earlier in Sect. 4.2.3, the acceptance conditions $\mathbf{AC}_{\text{prev}}(\alpha, \beta)$ and $\mathbf{AC}_{\text{next}}(\alpha, \beta)$ are standard in literature for $\alpha \in [1, 1.03]$ and $\beta \in [0, 0.2]$. Many agents tend to use these acceptance conditions, as they are well-known and easy to implement. We have formed Hypothesis 4.2 about them.

Hypothesis 4.2 $\mathbf{AC}_{\text{next}}(\alpha, \beta)$ will outperform $\mathbf{AC}_{\text{prev}}(\alpha, \beta)$ for all α and β. However, both conditions will perform worse than combined acceptance conditions, which also take the remaining time into account.

To test this hypothesis, we considered many different values for α and β in our experiments, with ranges chosen around the values we had found in existing agents (cf. Table 4.2).

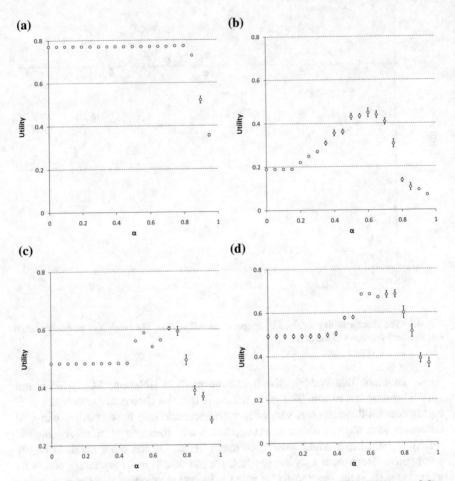

Fig. 4.2 The average utility of $\mathbf{AC}_{const}(\alpha)$ per negotiation scenario, for $\alpha \in [0, 1)$. **a** Laptop. **b** Itex versus cypress. **c** England–Zimbabwe. **d** Grocery

Consulting Table 4.5, the first observation is that $\mathbf{AC}_{prev}(\alpha, \beta)$ as well as $\mathbf{AC}_{next}$ (α, β) already perform much better than $\mathbf{AC}_{const}$ for all tested values of α and β. Higher values for α and β generally yield a better result, although the differences are quite small. However, given that we average the utility over 15 runs, we are able to statistically distinguish the performance for different values of α and β. We have found $\mathbf{AC}_{next}(\alpha, \beta)$ does indeed outperform $\mathbf{AC}_{prev}(\alpha, \beta)$ for all tested values of α and β, except for $\beta = 0$ (*two-tailed t-test*, $p < 0.01$), thereby partially confirming the hypothesis.

As an example, we have plotted $\mathbf{AC}_{next}(1, \beta) = \mathbf{AC}_{gap}(\beta)$ and $\mathbf{AC}_{next}(1, \beta)$ for $\beta \in [0, 1)$ in Fig. 4.3. We can confirm that $\mathbf{AC}_{next}(1, \beta)$ obtains scores that are significantly higher (using $p < 0.01$) scores than $\mathbf{AC}_{prev}(1, \beta)$, for $\beta \neq 0$.

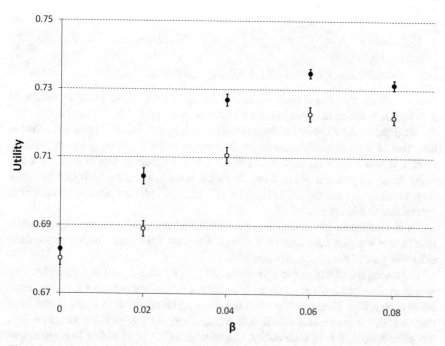

Fig. 4.3 The average utility obtained by agents using $\mathbf{AC}_{\text{next}}(1, \beta)$ (in *black*), and $\mathbf{AC}_{\text{prev}}(1, \beta)$ (in *white*). The *vertical error bars* indicate one standard deviation to the average utility of $N = 15$ different runs

It makes sense that comparing the opponent's offer to our upcoming offer is more beneficial than comparing it to our previous offer, as $\mathbf{AC}_{\text{next}}$ is always 'one step ahead' of $\mathbf{AC}_{\text{prev}}$. In general, $\mathbf{AC}_{\text{next}}$ is never worse than $\mathbf{AC}_{\text{prev}}$, and therefore there seems no reason to use the latter.

One of the top choices for both $\mathbf{AC}_{\text{next}}$ and $\mathbf{AC}_{\text{prev}}$, is setting $\alpha = 1.02$, and $\beta = 0.04$ (interestingly, *IAM(crazy)haggler* makes the same choice for α, cf. Table 4.2). However, even for this choice, the combined acceptance conditions $\mathbf{AC}_{\text{combi}}(T, \text{MAX}^W)$ outperform both of them for all tested values of T (*two-tailed t-test*, $p < 0.01$). This also settles the second part of the hypothesis.

The reason for the relatively bad performance of $\mathbf{AC}_{\text{next}}$ and $\mathbf{AC}_{\text{prev}}$ is that many bidding strategies focus on the 'negotiation dance' [17]. That is, modeling the opponent, trying to make equal concessions and so on. When a strategy does not explicitly take time considerations into account when making an offer, this poses a problem for these two standard acceptance conditions: they rely completely on the bidding strategy to concede to the opponent before the deadline occurs. When the agent or the opponent does not concede enough near the deadline, the standard conditions lead to poor performance.

Our third hypothesis with respect to the time-dependent condition is as follows:

Hypothesis 4.3 $\mathbf{AC}_{time}(T)$ always reaches an agreement, but of relatively low utility. This utility improves when T gets closer to the deadline.

To evaluate this hypothesis we tested $\mathbf{AC}_{time}(T)$ for many possible values of $T \in [0, 1)$, a selection of which can be examined in Table 4.5. We have found that the obtained utility increases monotonously with larger T, i.e.: it is optimal to choose the value of T sufficiently close to the deadline, while still allowing enough time to reach a win-win agreement. The fact that one has to accept as late as possible when using $\mathbf{AC}_{time}(T)$ clearly stems from the fact that we are dealing with undiscounted domains only; see our overall conclusions (Sect. 11.3.4) for a discussion on possible extensions in this regard.

From observing the acceptance probability of $\mathbf{AC}_{time}(T)$ in the experimental results, we see that the agent will always reach an agreement, therefore we consider this part of the hypothesis confirmed.

Regarding the utility of the agreement, $\mathbf{AC}_{time}(T)$ with $T < 1$ is a sensible criterion to avoid a break off at all costs. It is rational to prefer any outcome over a break off of zero utility. However, the resulting deal can be anything. As we can see from the table, this is the reverse situation of $\mathbf{AC}_{const}(0.9)$: $\mathbf{AC}_{const}(0.9)$ rarely gets a deal, but when it does, it is of high utility. Conversely, $\mathbf{AC}_{time}(T)$ yields a low agreement score (0.648 for $T = 0.95$), but with certainty of agreement. The overall score is the same (0.648), but it is interesting to note that this score is worse than all scores by either $\mathbf{AC}_{next}$ or $\mathbf{AC}_{prev}$ (*two-tailed t-test*, $p < 0.01$). This phenomenon can again be explained by the acceptance dilemma: by accepting any offer near the deadline, it reaches more agreements, but of relatively low utility.

This insight led us to believe that more consideration has to be given to the remaining time when deciding to accept an offer. That is why we conclude our analysis with combined acceptance conditions, which expand upon the idea to get better deals near the deadline.

When evaluating $\mathbf{AC}_{combi}(T, \alpha)$, we expected the following characteristics. First, $\mathbf{AC}_{combi}(T, \alpha)$ is an extension of $\mathbf{AC}_{next}$ in the sense that it will accept under broader circumstances. It alleviates some of the mentioned drawbacks of $\mathbf{AC}_{next}$ by also accepting when the utility gap between the parties is positive. Also note that in addition to the parameters that current acceptance conditions use, such as my previous bid $x_{A \to B}^{t_{n-1}}$, my next bid $x_{A \to B}^{t'}$, the remaining time, and the opponent's bid $x_{B \to A}^{t}$, this condition employs the entire bidding history $H_{A \leftrightarrow B}^{t}$ to compute the acceptability of an offer. Therefore we expect better results than with $\mathbf{AC}_{next}$, with more agreements, and when it agrees, we expect a better deal than by using $\mathbf{AC}_{time}(T)$.

We capture this last statement in our final hypothesis:

Hypothesis 4.4 The combination $AC_{combi}(T, \alpha)$ outperform other acceptance conditions, such as $AC_{time}(T)$ and $AC_{next}(\alpha, \beta)$, primarily by getting deals of higher utility.

As is evident from the experimental results, there are two acceptance conditions that dominate the others, namely $AC_{combi}(T, MAX^W)$, as well as $AC_{combi}(T, AVG^W)$ with T close to the deadline. The results are not statistically different for the different values of T, but any of the tested values performs quite well. One of the best AC's of the test is $AC_{combi}(0.98, MAX^W)$ with a score of 0.762, which is even better than the built-in mechanisms of the agents, and also surpasses the performance of $AC_{next}(\alpha, \beta)$ for any α and β (significantly so, using a *two-tailed t-test*, $p < 0.01$). In particular, it is at least 12 % better than AC_{next} (*two-tailed t-test*, $p < 0.01$).

Similar to AC_{time}, the combined conditions still get a deal almost every time, but with a higher payoff. However, the average utility of an agreement is not the highest: the built-in mechanisms and several $AC_{const}(\alpha)$ conditions get better agreements. But again, we can observe that their agreement rate is also lower, resulting in a higher overall score for the combined conditions. This settles our last hypothesis.

Finally, aiming for the highest utility that has been offered so far (i.e., using $AC_{combi}(T, MAX^T)$) is not as successful, mostly due to a big decrease in agreements. The higher utility that is obtained with this condition does not compensate for the loss of utility that is caused by a break off.

4.5 Conclusion

In this chapter, we aimed to classify current approaches to generic acceptance conditions and to compare a selection of acceptance conditions in a real-time setting. We presented the challenges and proposed new solutions for accepting offers in current state-of-the-art automated negotiations. The focus of this chapter is on decoupled acceptance conditions (i.e.: general conditions that do not depend on a particular bidding strategy), for which we have defined a formal negotiation model.

Designing an effective acceptance condition is challenging because of the acceptance dilemma: better offers may arrive in the future, but waiting for too long can result in a break off of the negotiation, which is undesirable for both parties.

We have presented and classified many of the standard acceptance conditions that are currently used by negotiating agents, including AC_{next}, AC_{prev}, and AC_{const}. From our results, it is apparent that they do not always yield optimal agreements, and we established that they perform worse than more sophisticated acceptance conditions.

In addition to classifying and comparing existing acceptance conditions, we have devised three new acceptance conditions by combining existing ones. This included two acceptance conditions that estimate whether a better offer might occur in the future based on recent bidding behavior. These conditions obtained the highest utility in our experiments and hence performed better than the other conditions we investigated. In particular, they outperform the acceptance mechanisms that are used by the top ANAC 2010 agents, and even the winner, *Agent K*, performs better when equipped with our combined acceptance conditions than with its built-in mechanism.

This chapter is based on the following publications: [18, 19]

Tim Baarslag, Koen V. Hindriks, and Catholijn M. Jonker. Effective acceptance conditions in real-time automated negotiation. *Decision Support Systems*, 60:68–77, Apr 2014

Tim Baarslag, Koen V. Hindriks, and Catholijn M. Jonker. Acceptance conditions in automated negotiation. In Takuyuki Ito, Minjie Zhang, Valentin Robu, and Tokuro Matsuo, editors, *Complex Automated Negotiations: Theories, Models, and Software Competitions*, volume 435 of *Studies in Computational Intelligence*, pages 95–111. Springer Berlin Heidelberg, 2013

References

1. van Galen Last N (2012) Agent Smith: opponent model estimation in bilateral multi-issue negotiation. In: Ito T, Zhang M, Robu V, Fatima S, Matsuo T (eds) New trends in agent-based complex automated negotiations. Studies in computational intelligence. Springer, Berlin, pp 167–174
2. Fatima SS, Wooldridge MJ, Jennings NR (2002) Optimal negotiation strategies for agents with incomplete information. In: Revised papers from the 8th international workshop on intelligent agents VIII, ATAL '01. Springer, London, pp 377–392
3. Faratin P, Sierra C, Jennings NR (1998) Negotiation decision functions for autonomous agents. Robot Auton Syst 24(3–4):159–182
4. Sierra C, Faratin P, Jennings NR (1997) A service-oriented negotiation model between autonomous agents. In: Boman M, van de Velde W (eds) Proceedings of the 8th European workshop on modelling autonomous agents in multi-agent world, MAAMAW-97, vol 1237 of Lecture Notes in Artificial Intelligence. Springer, pp 17–35
5. Fatima SS, Wooldridge MJ, Jennings NR (2002) Multi-issue negotiation under time constraints. In: AAMAS '02: proceedings of the first international joint conference on Autonomous agents and multiagent systems. ACM, New York, pp 143–150
6. Jonker CM, Robu V, Treur J (2007) An agent architecture for multi-attribute negotiation using incomplete preference information. Auton Agent Multi-Agent Syst 15:221–252
7. Hindriks KV, Tykhonov D (2008) Opponent modelling in automated multi-issue negotiation using bayesian learning. In: Proceedings of the 7th international joint conference on autonomous agents and multiagent systems, volume 1 of AAMAS '08, Richland, SC. International foundation for autonomous agents and multiagent systems, pp 331–338
8. Şerban LD, Silaghi GC, Litan CM (2012) AgentFSEGA—time constrained reasoning model for bilateral multi-issue negotiations. In: Ito T, Zhang M, Robu V, Fatima S, Matsuo T (eds) New trends in agent-based complex automated negotiations. Series of studies in computational intelligence. Springer, Berlin, pp 159–165
9. Williams CR, Robu V, Gerding EH, Jennings NR (2012) Iamhaggler: a negotiation agent for complex environments. In: Ito T, Zhang M, Robu V, Fatima S, Matsuo T (eds) New trends in agent-based complex automated negotiations. Studies in computational intelligence. Springer, Berlin, pp 151–158
10. Kawaguchi S, Fujita K, Ito T (2012) Compromising strategy based on estimated maximum utility for automated negotiating agents. In: Ito T, Zhang M, Robu V, Fatima S, Matsuo T (eds) New trends in agent-based complex automated negotiations. Series of studies in computational intelligence. Springer, Berlin, pp 137–144

11. An B, Lesser VR (2012) Yushu: a heuristic-based agent for automated negotiating competition. In: Ito T, Zhang M, Robu V, Fatima S, Matsuo Tokuro (eds) New trends in agent-based complex automated negotiations, vol 383., Studies in computational intelligenceSpringer, Berlin, pp 145–149

12. Faratin P, Sierra C, Jennings NR (2002) Using similarity criteria to make issue trade-offs in automated negotiations. Artif Intell 142(2):205–237

13. Hindriks KV, Tykhonov D (2010) Towards a quality assessment method for learning preference profiles in negotiation. In: Ketter W, La Poutré JA, Sadeh N, Shehory O, Walsh W (eds) Agent-mediated electronic commerce and trading agent design and analysis. Lecture notes in business information processing, vol 44. Springer, Berlin, pp 46–59

14. Kersten GE, Zhang G (2003) Mining inspire data for the determinants of successful internet negotiations. Central Eur J Oper Res 11(3):297–316

15. Lin R, Kraus S, Tykhonov D, Hindriks KV, Jonker CM (2011) Supporting the design of general automated negotiators. In: Proceedings of the second international workshop on agent-based complex automated negotiations (ACAN'09), vol 319. Springer, Berlin, pp 69–87

16. Lin R, Kraus S, Wilkenfeld J, Barry J (2008) Negotiating with bounded rational agents in environments with incomplete information using an automated agent. Artif Intell 172(6–7):823–851

17. Raiffa H (1982) The art and science of negotiation: How to resolve conflicts and get the best out of bargaining. Harvard University Press, Cambridge

18. Baarslag T, Hindriks KV, Jonker CM (2014) Effective acceptance conditions in real-time automated negotiation. Decis Support Syst 60:68–77

19. Baarslag T, Hindriks KV, Jonker CM (2013) Acceptance conditions in automated negotiation. In: Ito T, Zhang M, Robu V, Matsuo T (eds) Complex automated negotiations: theories, models, and software competitions. Studies in computational intelligence, vol. 435. Springer, Berlin, Heidelberg, pp 95–111

Chapter 5
Accepting Optimally with Incomplete Information

Abstract In the previous chapter we classified generic acceptance conditions, and we formulated new ones that performed better in an experimental setting. This chapter takes a different approach by devising theoretically *optimal* solutions. We approach the decision of whether to accept as a *sequential decision problem*, by modeling the bids received as a stochastic process. We argue that this is a natural choice in the context of a negotiation with incomplete information, where the future behavior of the opponent is uncertain. We determine the *optimal acceptance policies* for particular opponent classes and we present an approach to estimate the expected range of offers when the type of opponent is unknown. We apply our method against a wide range of opponents, and compare its performance with acceptance mechanisms of state-of-the-art negotiation strategies. The experiments show that the proposed approach is able to find the optimal time to accept, and improves upon widely used existing acceptance mechanisms.

5.1 Introduction

Suppose two parties A and B are conducting a negotiation, and B has just proposed an offer to A. A is now faced with a decision: she must decide whether to continue, or to accept the offer that is currently on the table. On the one hand, accepting the offer and ending the negotiation means running the risk of missing out on a better deal in the future. On the other hand, carrying on with the negotiation involves a risk as well, as this gives up the possibility of accepting one of the previous offers. How then, should A decide whether to end or to continue the negotiation?

Of course, A's decision making process will depend on the current offer, as well as the offers that A can expect to receive from B in the future. However, in most realistic cases, agents have only incomplete information about each other [5, 9, 15]. In this chapter, we explore in particular the setting where the opponent has only limited or no knowledge of A's preferences, and the proposals that A receives will therefore be necessarily uncertain. This makes A's task of predicting B's future offers by no means an easy one.

© Springer International Publishing Switzerland 2016
T. Baarslag, *Exploring the Strategy Space of Negotiating Agents*,
Springer Theses, DOI 10.1007/978-3-319-28243-5_5

Moreover, predicting B's future offers is only part of the solution: even when A can predict B's moves reasonably well, A still has to decide how to put this information to good use. In other words, even when a probability distribution over the opponent's actions is known, it is not straightforward to translate this into effective negotiation behavior. As an extreme example, consider an opponent R (for Random) who will make random offers with utility uniformly distributed in $[0, 1]$. Suppose furthermore that we can expect to receive two more bids from R until the deadline occurs. R currently makes an offer of utility $x \in [0, 1]$; for what x should we accept? Of course, an even better bid than x might come up in one of the two remaining rounds; on the other hand, it might be safer to settle for this bid if x is large enough. For this particular case, we will prove that there is an optimal acceptance strategy, and we show exactly for what x to accept (see Sect. 5.3).

The main contribution of this chapter is that we address both of A's problems: first, at every stage of the negotiation, we provide a technique to estimate the bidding behavior of various opponent classes by modeling A's dilemma as a stochastic decision problem. For particular opponent classes we are able to provide precise models, and to formulate exact mathematical solutions to our problem. For the second step, using the ranges found earlier, we borrow techniques from optimal stopping theory to find generic, optimal rules for when to accept against a variety of opponents in a bilateral negotiation setting with incomplete information. The solutions proposed are optimal in the sense that there can be no better strategy in terms of utility.

We begin by introducing our approach in Sect. 5.2, and we apply our methods to find optimal rules in the specific case of opponents that bid randomly in Sect. 5.3. We then build upon these cases and subsequently work out more realistic scenarios in the following sections. In Sect. 5.4, we explore opponents that change their behavior over time, and we determine optimal stopping rules when good estimates of their bidding behavior are known. We extend these results by combining our approach with a state-of-the-art prediction mechanism, and we demonstrate that our approach outperforms existing accepting mechanisms, even when the opponent's behavior is unknown.

5.2   Decision Making in Negotiation Under Uncertainty

As defined in Sect. 2.2.2, we focus on a bilateral negotiation, wherein two agents try to reach an agreement while maximizing their own utility using the *alternating-offers protocol*. However, instead of a real time line, the negotiating parties take turns in exchanging offers for a *fixed number of rounds N*. In case this deadline is reached before both parties come to an agreement, both receive zero utility. A preference profile is described by a utility function $u(x)$, which maps each possible outcome x in the negotiation domain to a utility in the range $[0, 1]$. If an agent receives an offer x, its acceptance mechanism has to decide whether $u(x)$ is high enough to be acceptable.

5.2.1 Stochastic Behavior in Negotiation

Suppose player B is involved in a bilateral negotiation with private preference information, and at some point in time, he has decided that he is satisfied with a utility around u, called his *target utility*. However, there are many possible bids with approximately this utility for player B. As usual, we call this set of bids X the *iso-level* bids with utility u. As player B is indifferent between these bids, B may attempt to optimize player A's utility in order to maximize the chance of an agreement. But this is difficult for B to achieve, as he does not know A's preferences. When using the alternating offers protocol, player B cannot simply send out all considered offers as one bundle, but instead, he can only offer them sequentially. Player B typically continues to select different bids from X until his target utility changes; then a new set of iso-level bids is generated, and the process starts again. The order in which player B picks bids from the set of equally preferred bids X will differ per player, but due to incomplete information, he can only select a bid with a particular opponent utility with limited certainty. Therefore, we can reasonably model the offers that are presented to A as a *stochastic process*.

This kind of stochastic behavior can be observed in practice in the Automated Negotiating Agents Competition (Appendix B). ANAC is a yearly international competition in which negotiation agents compete in an incomplete information setting. Half of the participants of ANAC 2011 [1, 3, 7, 13, 24] were not designed to explicitly optimize opponent utility and therefore, with the limited information available, simply selected a random element from X; others used opponent modeling techniques that estimate the opponent's preferences in order to select bids closer to the Pareto optimal frontier. However, opponent modeling is seldom capable of making perfect estimates [19]. Consequently, even when player B employs an opponent modeling technique, A will still receive bids of varying utility. Moreover, the agents usually already anticipated the limitations of their opponent model, and therefore randomly chose among the estimated top bids for the opponent [12, 23], adding even further to the random appearance of the utility of their bids. As a result, the negotiation traces of ANAC 2011 showed to a very large extent the stochastic behavior discussed above (see also Fig. 5.1). Only 25 % of the negotiation moves were an improvement for the opponent over the previous bid; the other 75 % of the moves could be classified as selfish, unfortunate, or silent [4].

In the next paragraph, we present our model of bid reception as a stochastic process, and then present optimal stopping techniques to optimize A's expected utility.

5.2.2 Optimal Stopping in Negotiation

In 1613, the celebrated astronomer Johannes Kepler wrote a long letter to Strahlendorf in which he describes a great problem that he faced [10, 11]. Kepler had lost his wife and set about finding a new wife through a series of interviews among eleven

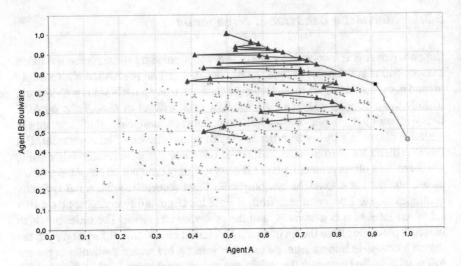

Fig. 5.1 Despite the fact that Side *B* concedes predictably over time, the utility of the offers seem to be randomly distributed around the [0.4, 0.8] interval for Side *A*, and as a result, the best bids for *A* occur during the middle of the negotiation

candidates. In order not to hurt the feelings of his potential wives, he would have to interview them sequentially and make a decision to marry them before moving on to the next candidate. His problem was: how do I decide to stop looking and settle for one of the candidates? Kepler's problem is now known as an instance of an *optimal stopping problem*: a stochastic decision problem of determining whether to accept among offers appearing sequentially and randomly.

We can frame the problem of accepting a bid as an optimal stopping problem [6], in which an agent is faced with the dilemma of choosing when to take a particular action, in order to maximize an expected reward or minimize an expected cost. In such problems, observations are taken sequentially, and at each stage, one either chooses to stop to collect, or to continue and take the next observation (usually at some specified sampling cost).

The model of bid reception is as follows: at each of a total of *N* rounds, we receive a bid, which has an associated utility, or value, drawn from a random variable over the unit interval. At this point, we must decide whether to accept the bid, or not. Once we accept, the deal is settled and the negotiation ends. If we continue, then there is no possibility of recalling passed-up offers; i.e., previous offers are unavailable unless they are presented to us again. Hence, at each round, we must decide to either continue or to stop participating in the negotiation, and we wish to act so as to maximize the expected net gain. Once an offer is turned down, and we decide to wait for another bid (at a cost *C*), the total number of remaining observations decreases by one. We will develop the theory here for arbitrary sampling cost *C*, but in the remainder of the thesis, we will assume the cost to be zero.

At every stage, the current situation may be described by a *state* (j, x), which is characterized by two parameters: the number of remaining observations $j \in \mathbb{N}$, and the latest received offer $x \in [0, 1]$. Let the utility distribution with j rounds remaining be given by a random variable X_j, with associated distribution function F_j. We can think of X_j as the possible utilities we receive when the opponent makes iso-level bids, and $F_j(u)$ represents the probability of receiving a bid with utility less than or equal to u. The expected payoff is then given by

$$V(j, x) = \max(x, E(V(j - 1, X_{j-1})) - C), \qquad (5.1)$$

where we abbreviate the second term $E(V(j - 1, X_{j-1})) - C$ as v_j. This represents the expected value of rejecting the offer at (j, x), and going on for (at least) one more period. Note that v_j does not depend on x. Thus, using the substitution, we get

$$v_j = E(\max(X_{j-1}, E(V(j - 2, X_{j-2})) - C) - C,$$

which leads to the following recurrence relation:

$$\begin{cases} v_0 = 0, \\ v_j = E(\max(X_{j-1}, v_{j-1})) - C. \end{cases} \qquad (5.2)$$

In [6, 16] it is proven that for any $s \in \mathbb{R}$, and for any random variable X with distribution function F for which $E(X)$ is finite, the following holds:

$$E(\max(X, s)) = s + T_F(s),$$

with

$$T_F(s) = \int_s^\infty (1 - F(t))\, dt.$$

And therefore the recurrence relation describing v_j can be written as

$$v_j = v_{j-1} + \int_{v_{j-1}}^\infty (1 - F_{j-1}(t))\, dt - C. \qquad (5.3)$$

Thus, if we know the distribution F_j for every j, we can compute the values v_j using the above recurrence relation. Then, deciding whether to accept an offer x is simple: if $x \geq v_j$ we accept, otherwise we reject the offer (see Algorithm 1).

There is however, a serious impediment to using our stochastic decision model in practice: we do not *know* the distributions of the utility that the opponent will present to us in the upcoming rounds; furthermore, the distributions are highly influenced by the specifics of the negotiation scenario.

However, against specific classes of opponents, we are able to establish these probabilities, and in an exact way. We will first focus our attention on theoretical

cases that resemble the relevant cases encountered in practice. In order to develop
the theory, we will first take on the extreme case of random opponent behavior, and
gradually add complexity as we proceed.

Of course, in a general setting we do not know the opponent's behavior, and in
that case we require a method to determine the distributions X_j for every remaining
round j. This means that for every round, we need to estimate the probability of
receiving certain utility in *our* utility space. This is the most difficult case, which we
will cover at the end of the chapter.

All in all, we consider three different opponent classes:

1. Random behavior: *fixed* and *known* uniform X_j in every round; this is solved
   mathematically in Sect. 5.3.
2. Known time-dependent behavior: *changing*, but *known* uniform X_j; this is optimally solved in Sect. 5.4.1.
3. Unknown time-dependent behavior: *changing* and *unknown* arbitrary X_j; covered
   in Sect. 5.4.2.

We will start with the first case, where we consider random opponent behavior.

5.3 Accepting Random Offers

Suppose an agent A is negotiating with its opponent, and the deadline is approaching,
so both agents have only a few more offers to exchange. As argued above, the opponent will often offer bids with varying utility for A, due to its incomplete information
of what A exactly wants. This means that from A's point of view, the utility of the
presented offers will have a particular stochastic distribution. The aim of A is then
to pick the best one given the limited time that is left.

We start by studying the extreme case of a maximally unpredictable opponent, or
Random Walker (see Sect. 2.3.3), who makes random bids at every stage of the negotiation. We first solve this case analytically, before moving on to more complicated
settings.

There are two crucial properties of *Random Walker* that make it a simplified case:
first, it picks random bids uniformly from the bid space; second, it is stateless; i.e.,
it uses the same decision function in every round of the negotiation, regardless of
the behavior of the other party or the time that has passed. We will weaken both
constraints in later stages of this chapter.

5.3.1 Uniformly Random Behavior

Using Eq. (5.2), we can determine the optimal solution against *Random Walker*,
using the added knowledge that every X_j does not depend on the number of rounds
left, and assuming every X_j is uniformly distributed:

Proposition 5.1 *Against an opponent who makes random bids of utility uniformly distributed in* $[0, 1]$*, and with* j *offers still to be observed, one should accept an offer of utility* x *exactly when* $x \geq v_j$*, where* v_j *satisfies the following equation:*

$$\begin{cases} v_0 = 0, \\ v_j = \frac{1}{2} + \frac{1}{2}v_{j-1}^2, \end{cases} \tag{5.4}$$

This recurrence relation has the following properties: v_j *is monotonically increasing, and*

$$\lim_{j \to \infty} v_j = 1.$$

Proof Let X be the uniform distribution over $[0, 1]$ with distribution function F. Playing against *Random Walker*, all X_j's are uniform distributions over $[0, 1]$ and hence equal to X. This yields:

$$T_F(s) = \int_s^\infty (1 - F(t)) \, dt \tag{5.5}$$

$$= \begin{cases} \frac{1}{2} - s, & s < 0. \\ \frac{1}{2}(1 - s)^2, & 0 \leq s \leq 1. \\ 0, & s > 1. \end{cases} \tag{5.6}$$

Since we are in the case $0 \leq s \leq 1$, we get:

$$v_j = v_{j-1} + \frac{1}{2}(1 - v_{j-1})^2 = \frac{1}{2} + \frac{1}{2}v_{j-1}^2.$$

It is easy to show with recursion that

$$1 - \frac{2}{j} < v_{j-1} < v_j < 1,$$

and therefore,

$$\lim_{j \to \infty} v_j = 1.$$

When we substitute $v_j = 1 - 2x_j$ in Eq. (5.4), we get the equivalent relation of the logistic map $x_j = x_{j-1}(1 - x_{j-1})$ at $r = 1$, which due to its chaotic behavior does not in general have an analytical solution. However, we have visualized its behavior for $j \in [0, 200]$ in Fig. 5.3 (uniform case). From this, we see that the answer to the question posed in the introduction is as follows: with two rounds to go, one should accept an offer x exactly when $x \geq v_2 = 0.625$.

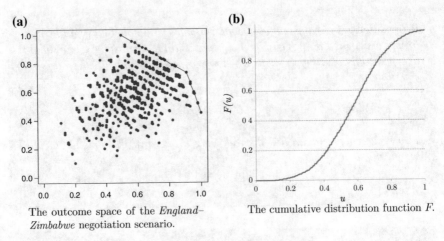

The outcome space of the *England–Zimbabwe* negotiation scenario.

The cumulative distribution function F.

Fig. 5.2 The outcome space for player A and B, and the resulting cumulative distribution function for player A

Note that $\lim_{j \to \infty} v_j = 1$ means we can expect to receive utility arbitrarily close to the maximum given enough time, and that this limit $v = 1$ is also the fixpoint of recurrence relation (5.4).

5.3.2   Non-Uniform Random Behavior

In Proposition 5.1, we consider random behavior by the opponent in a uniform way; i.e., a scenario where every received utility is equally likely. However, in practice such situations rarely occur. Negotiation scenarios usually enable agents to make trade-offs between multiple issues, resulting in clustering of potential outcomes. Hence, in a typical scenario, even when the opponent chooses bids randomly, the utilities of those bids are *not* distributed uniformly.

A typical example of such a multi-issue negotiation scenario is depicted in (see Fig. 5.2a), and involves a case where *England* and *Zimbabwe* negotiate an agreement on tobacco control (see Sect. C.2 of the Appendix). The leaders of both countries have contradictory preferences for two issues, but three other issues have options that are jointly preferred by both sides. We will use it as a running example, but the outlined technique can be applied to any negotiation scenario.

For such an outcome space, we cannot simplify Eq. (5.5) any further, so instead, we need to integrate the cumulative distribution function $F(x)$ directly (see Fig. 5.2b). Note that $F(x)$ can be computed by the agent, simply by considering the distribution of the utilities of all possible outcomes. Using Eq. (5.3), we can now compute the values v_j for a scenario such as *England–Zimbabwe*; see Fig. 5.3.

j	Unif. v_j	E-Z v_j
0	0	0
1	0.5	0.5734
2	0.625	0.6449
3	0.6953	0.6855
4	0.7417	0.7134
5	0.7751	0.7344
10	0.8611	0.7953
100	0.9812	0.9338
200	0.9903	0.9586

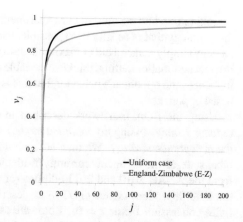

Fig. 5.3 The optimal stopping values v_j for different rounds j versus uniformly random (*Uniform case*) and non-uniformly random (*E–Z*) behavior

Note that the value of v_j for *England–Zimbabwe* increases faster than in the uniform case, but at the same time it also tends to 1 more slowly. This can be explained by the fact that this outcome space is more sparse in both extremes: since there are less bids of very low utility, it should aim higher at the end of the negotiation, and as there are also less bids of high utility, it should be satisfied more easily at the start of the negotiation.

5.3.3 Experiments

In order to test the efficacy of the optimal stopping condition, we first integrated it into a functional negotiating agent. This requires care, as normally, the behavior (and thus the performance) of a negotiating agent is determined by many factors outside of the acceptance mechanism, particularly its bidding strategy. Note however, that against *Random Walker*, bidding strategies with the same acceptance policy perform equally, as it does not matter which offers are sent out. This holds because of three properties:

1. *Random Walker*'s offers do not depend on the opponent's behavior; hence, it is not sensitive to the other's bidding strategy;
2. *Random Walker* does not accept any offers; in our experiments, opponents are not allowed to accept, as this could prematurely end the negotiation, without revealing anything about the performance of the acceptance strategies.
3. The optimal stopping condition works independently of bids that are sent out.

Taking this into account in our experiments, we opted for an accompanying bidding strategy that is as simple as possible, namely *Hardliner* (see Sect. 2.3.3). This strategy simply makes a bid of maximum utility for itself and never concedes. Clearly, in a real negotiation setting, this is not a viable bidding tactic as it generally negatively influences the opponent's behavior, but this is of no concern against a non-behavior-based opponent.

Using the BOA framework described in Chap. 3, we combined the *Hardliner bidding strategy* (using no *opponent model*) with the optimal stopping condition as the *acceptance strategy*. We then compared its performance with the strategies of other state-of-the-art agents currently available for our setting. We selected all agents from the ANAC 2010 and 2011 editions (cf. Sect. C.1 and D.1). We also included the time-dependent tactics (TDT's) as described in Sect. 2.3.3, such as *Hardliner* (with concession factor $e = 0$), *Boulware* ($e = \frac{1}{2}$), *Conceder Linear* ($e = 1$), and *Conceder* ($e = 2$) taken from [8]. To analyze the performance of different agents, we employed our GENIUS environment [18] (cf. Appendix A).

Algorithm 1: Optimal Stopping main decision body

Input: The number of remaining rounds j, and the last received bid x by the
        opponent.
Output: Acceptance or rejection of x.
begin
    offeredUtility $\longleftarrow$ getUtility(x);
    target $\longleftarrow$ determine(v_j);
    if offeredUtility $\geq$ target **then**
        $\llcorner$ **return** ACCEPT
    else
        $\llcorner$ **return** REJECT // And send a counter-offer

---

For our negotiation scenarios, we opted for the *England–Zimbabwe* domain described in Sect. C.2 of the Appendix, and a discretized version of *Split the Pie* [20, 21], where two players have to reach an agreement on the partition of a pie of size 1. The pie will be partitioned only after the players reach an agreement, in which case one gets $x \in [0, 1]$, and the other gets $1 - x$. In this scenario, *Random Walker* makes bids of utility uniformly distributed in [0, 1], since it proposes random partitions of the pie.

The results of our experiment in the uniform case and on *England–Zimbabwe* are plotted in Figs. 5.4 and 5.5 respectively, both for $N = 10$ and $N = 100$ negotiation rounds. The optimal stopping condition significantly outperforms all agents (*one-tailed t-test, $p < 0.01$*) in all cases.

In the uniform case (cf. Fig. 5.4), it obtains the highest score possible both in 10 and in 100 rounds, getting respectively 86 and 98 % of the pie on average. Note that this is exactly equal to the theoretical values v_{10} and v_{100} shown in the uniform case of Fig. 5.3.

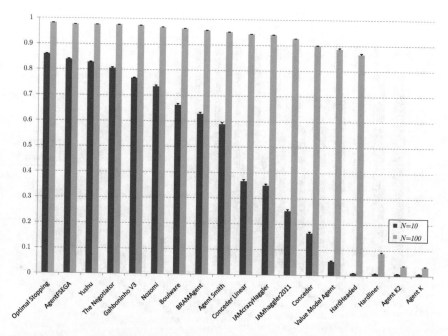

Fig. 5.4 The optimal stopping condition outperforms all ANAC agents against uniform *Random Walker* for 10 and 100 rounds. Average utility plotted over 5000 runs; the *errors bars* indicate one standard deviation to the mean

On *England–Zimbabwe* (cf. Fig. 5.5), the optimal stopper obtains less utility for $N = 10$ in an absolute sense compared to the uniform case, but the results are even more pronounced relative to the other agents. A moment of reflection makes clear why: given the clustering of bids of medium utility (see Fig. 5.2a), there is less chance for *Random Walker* to propose a very fortunate bid for the opponent. This explains why the acceptance strategies of the other agents perform relatively worse. Note that the end result obtained by the optimal stopper is approximately 0.79 and 0.93 for $N = 10$ and $N = 100$ respectively, which is again equal to the optimal values v_{10} and v_{100} shown in the E–Z case of Fig. 5.3.

5.3.4 When Optimal Stopping Is Most Effective

As is evident from the results, optimal stopping performs better than the other agents against *Random Walker*. This is to be expected given the fact that no agent could possibly do better; however, the difference with the current state-of-the-art is surprisingly big in some cases, for example compared to the ANAC 2010 winner, *Agent K* [13], and even more so when the number of rounds is limited. The reason is that

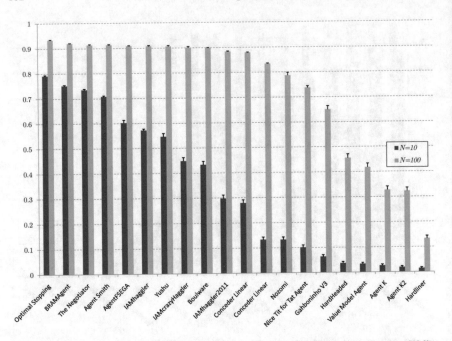

Fig. 5.5 The optimal stopping condition outperforms all ANAC agents against *Random Walker* on *England–Zimbabwe* for $N = 10$ and $N = 100$ rounds; utility averaged over 5000 runs

many of the currently used acceptance mechanisms are rather straightforward, and only become successful against *Random Walker* when enough time is available.

This can be illustrated by considering the baseline acceptance condition where an agent accepts if and only if the offered utility is above a fixed threshold α, as is done by various agents [13, 22, 24, 25]. In the uniform case, its expected utility obtained over N rounds equals the probability it will obtain an offer above α multiplied by the expected utility above α:

$$(1 - \alpha^N) \cdot \frac{\alpha + 1}{2}. \tag{5.7}$$

This is not a very efficient acceptance condition for small N; for example, for $N = 2$, the optimal value of α (i.e., the value of α that maximizes formula (5.7)) is $\frac{1}{3}$, with expected utility of 0.593, while the optimal value that can be obtained is $v_2 = 0.625$. However, for large N, choosing α close to one is already surprisingly efficient. For example, for $N = 100$, the optimal value of α is 0.948, with expected utility of 0.969. This is already quite close to the optimal value of $v_{100} = 0.981$; this indicates that in case bids are randomly distributed, the added value of our solution lies primarily in negotiations with a limited amount of total rounds, or when only limited time is left in a negotiation. Therefore, in the next sections, we extend our results in order to tackle more generic circumstances. Some earlier solutions can be derived from the more general ones treated later; however, we have elected to go from simple to the more complicated.

5.4 Time Dependent Offers

One of the most restrictive assumptions so far was to assume the opponent plays completely randomly. As we argued, this is a sensible assumption when modeling an opponent that is extremely unpredictable due to imperfect information, but the general case is more complicated. Almost all negotiation agents change their range of offers over time; i.e., they are *time dependent strategies*.[1] Hence, we require optimal stopping policies for these cases as well.

The challenge of the more general case is that we have to account for the fact that not only the presented utilities may fluctuate, but also the *range* of future offers may be different at different times. Establishing this range is not easy, because the strategy used by the opponent is of course unknown to us. The offers of any time dependent opponent with incomplete information can again be modeled by a stochastic distribution, but this time the distribution will change over time. In terms of optimal stopping, this means that the bid distribution X_j can be different for every j.

5.4.1 Uniformly Unpredictable Offers

If we assume that the opponent's offers are uniformly distributed, we only need to know the *interval* of utilities we can expect in every round. If this is the case, then we are able to compute the optimal time to accept, as is stated in the following proposition.

Proposition 5.2 *Against a time-dependent opponent who, with j rounds still to be observed, makes bids uniformly distributed in $X_j = [a_j, b_j]$, the optimal stopping cut-off is v_j, where v_j satisfies the following equation:*

$$v_j = \begin{cases} 0, & \text{if } j = 0 \\ v_{j-1}, & \text{if } v_{j-1} \geq b_{j-1} \\ \frac{b_{j-1}+a_{j-1}}{2}, & \text{if } v_{j-1} \leq a_{j-1} \\ v_{j-1} + \frac{1}{2} \cdot \frac{(b_{j-1}-v_{j-1})^2}{b_{j-1}-a_{j-1}}, & \text{if } a_{j-1} < v_{j-1} < b_{j-1}. \end{cases}$$

---

[1]Note that these should not be confused with the well-known time-dependent *tactics* as described in Sect. 2.3.3, which are particular kinds of time dependent strategies.

Proof From Eq. (5.2), we have

$$v_j = E(\max(X_{j-1}, v_{j-1})),$$

so immediately, if $v_{j-1} \geq b_{j-1}$, then $v_{j-1} \geq X_{j-1}$, and thus $v_j = v_{j-1}$. On the other hand, if $v_{j-1} \leq a_{j-1}$, then $v_j = E(X) = \frac{b_{j-1}+a_{j-1}}{2}$. So therefore, the only case left is $a_{j-1} < v_{j-1} < b_{j-1}$, in which case we derive the following:

$$v_j = P(X_{j-1} \leq v_{j-1}) \cdot v_{j-1} + P(X_{j-1} > v_{j-1}) \cdot P(X_{j-1} \mid X_{j-1} > v_{j-1})$$

$$= \frac{v_{j-1} - a_{j-1}}{b_{j-1} - a_{j-1}} \cdot v_{j-1} + \frac{1}{2} \cdot \frac{b_{j-1} - v_{j-1}}{b_{j-1} - a_{j-1}} \cdot (v_{j-1} + b_{j-1})$$

$$= v_{j-1} + \frac{1}{2} \cdot \frac{(b_{j-1} - v_{j-1})^2}{b_{j-1} - a_{j-1}}.$$

Note how this proposition is an extension of Proposition 5.1: if we set $X_j = [0, 1]$ for every j, the equation simplifies to $v_j = \frac{1}{2} + \frac{1}{2}v_{j-1}^2$ again.

Also, we observe that in the special case of perfect information, the distributions would be singletons of the form $X_j = \{x_j\}$, with probability 1 for the outcome x_j. The equation of Proposition 5.2 then simplifies to

$$v_j = \begin{cases} 0, & \text{if } j = 0 \\ \max(x_{j-1}, v_{j-1}), & \text{otherwise.} \end{cases}$$

$$= \max_{0 \leq k < j} x_k.$$

This means that the optimal stopping procedure has the desirable property that when it gets perfect estimates as input, it will also produce perfect output.

5.4.2 Arbitrarily Unpredictable Offers

Proposition 5.2 is useful to gain insight into the optimal acceptance policy, but in practice, the distributions X_j are neither known, nor uniformly distributed, and therefore an estimation method is required against arbitrary opponents. Of course, the success of the optimal stopping rules will greatly depend on the fidelity of the estimating technique used to predict the opponent's behavior. Therefore, we first examine the case of a perfect estimator, to see how our method performs in the ideal case. After that, we will move our focus to an estimator that can be used in practice.

5.4.2.1   Opponent Prediction Using Perfect Estimates

The perfect estimation method that we employ divides the number of rounds N into a number of time slots S. Then, by momentarily using perfect information, it gets the minimum and maximum utility that will be offered by the opponent during that time slot. This allows us to control exactly the precision of the estimate, where using more slots emulates having more information about the opponent's behavior. If we set the number of slots equal to the total number of rounds, we are in a full information state and the performance should be theoretically optimal. If we use only one slot, we have less information, knowing only the opponent's utility range over the entire N rounds.

5.4.2.2   Opponent Prediction Using Gaussian Process Regression

Finally, we consider an estimation method that uses as input only the information that can be observed during the negotiation, namely the utility of the offers made by the opponent. For this, we opted for a Gaussian process regression (GPR) technique as described in [24]. We selected the GPR technique because it can be computed in real time during the negotiation, and it is specifically designed to be robust with respect to significantly varying observations. It works as follows: for each offer made by the opponent, the round at which the offer was made is recorded, along with the offered utility. From this, the future concessions of the opponent are estimated using regression with a Gaussian process. To reduce the effect of noise, the offers received are aggregated in a number of time windows, and only the maximum value that is received in each time window is used as input for the Gaussian process.

The output of the Gaussian process regression is a normal distribution for every upcoming round k, with mean μ_k and standard deviation σ_k. The mean μ_k gives a prediction of the most likely offered utility value in round k, whilst the standard deviation σ_k gives an indication of how accurate the prediction is. When using GPR, the opponent bid distribution is estimated in real-time by a *normal* distribution, truncated to fit in the range [0, 1].

Algorithm 2: Determining v_j

Input: The number of remaining rounds j, and all negotiation outcomes Ω.
Output: v_j.
begin

    if $j = 0$ **then**
        └ **return** 0
    // Use either perfect estimation, or GPR
    $Y_{j-1} \longleftarrow$ estimated utility distribution at $j - 1$;
    // Use either uniform, or Gaussian distribution for X_{j-1}
    $X_{j-1} \longleftarrow$ utility distribution of Y_{j-1} over Ω;
    // Recursively determine v_{j-1}
    return $E(\max(X_{j-1}, v_{j-1}))$

Table 5.1 Characteristics of the negotiation scenarios used in the experiments

Scenario	Size	Opposition
Car	15625	Low
Amsterdam	3024	Medium
England–Zimbabwe [17]	527	Medium
Nice Or Die [3]	3	High
Itex versus Cypress [14]	180	High
Travel	188160	Medium

5.4.3 Experiments

To analyze the performance of optimal stopping (OS) against time dependent negotiation strategies, we adopted the same experimental setup as before, this time testing it with both perfect estimates and the GPR technique. We set the number of Gaussian process regressions to 10, and we set the number of samples equal to the number of rounds (for details, see [24]).

We used two versions of perfect estimation: the full information state by setting $S = N$ (called "perfect estimation with full slots"), and a version with $S = 1$ (called "perfect estimation with one slot"). We also tested two variants of the GPR technique: in one, we simply set X_j equal to the truncated Gaussian distribution with mean μ_j and standard deviation σ_j as predicted by the GPR technique (called "Gaussian GPR prediction"). These predictions turned out to be overly optimistic in most cases, since the GPR technique uses as input the maximum utility received in each time slot. Therefore, we opted to include a simplified second version, which produces a uniform distribution between zero and the estimated maximum offered utility, which we set to $\mu + 2\sigma$ (called "uniform GPR prediction"). See also Algorithm 2.

As the specifics of the negotiation scenario influences the behavior of the opponent, we picked a total of six negotiation scenarios from ANAC 2010 and 2011 (see Appendix C and D), aiming for a large spread of negotiation characteristics (see Table 5.1).

For the opponents, we selected various TDT's from [8], as defined in Sect. 2.3.3. Our optimal stopping policy works against *any* type of time dependent negotiation strategy, but we selected TDT's because they are typical, well-known examples of strategies that change their range of bids over time. Additionally, as in the case of *Random Walker*, TDT's are non-adaptive and hence it is not important what counter-offers are sent out to them.

We selected the same TDT's used earlier, namely: *Hardliner*, *Boulware*, *Conceder Linear*, and *Conceder*. We generated many variants of each opponent by choosing the values 0.7, 0.8, and 0.9 for the *min* parameter (which controls the utility threshold up to where the agent will concede [8]). Note that this creates quite a competitive

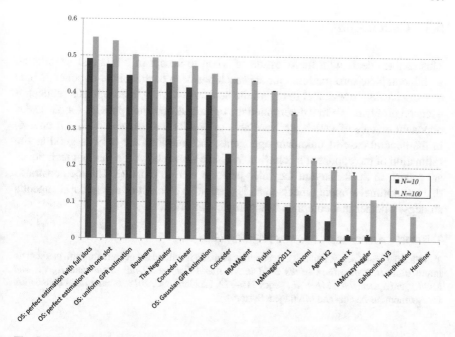

Fig. 5.6 The utilities obtained by ANAC agents and optimal stopping conditions with different estimation methods against time dependent opponents for 10 and 100 rounds

opponent pool, as the opponents will never fully concede. This leads to only small utility differences between the different acceptance strategies, but this should be regarded an artifact of our competitive setup. The average score of all agents is shown in Fig. 5.6.

Optimal stopping with perfect estimation with full slots should be considered the theoretical upper bound here; and indeed, it outperforms all other methods. Among the agents that act with *incomplete* information, the *Boulware* agent obtains a surprisingly good score. Its strategy turns out to be particularly successful against TDT's since it waits for a long time to let the opponent concede as much as possible, until it quickly concedes in the end to obtain an agreement. However, because it waits for so long, it misses out on good offers that are offered earlier.

Gaussian GPR prediction is not as successful, mainly because it was found to overestimate the opponent's willingness to concede, and hence it aimed for too much during the negotiation. It is optimal stopping with uniform GPR that performs significantly best (*one-tailed t-test*, $p < 0.01$), which shows that the optimal stopping policy is indeed a robust mechanism that can still perform well in an incomplete information setting.

5.5 Conclusion

This chapter deals with the question of when to accept in a bilateral negotiation with incomplete information. Our approach has been to model the opponent's bids as a stochastic process, and to regard the decision of when to accept as a sequential decision problem. We first determined the optimal acceptance policies for particular opponent classes of which we were able to predict the behavior well. Of course, in the general case of unknown opponents, the solutions are only as good as the estimation of the opponent's behavior. We have shown however, that our techniques are robust, in the sense that they also perform well in practice. This demonstrates that our optimal stopping mechanism is a valuable element of a negotiating agent's strategy, whether in a complete or incomplete information setting.

This chapter is based on the following publication: [2]

Tim Baarslag and Koen V. Hindriks. Accepting optimally in automated negotiation with incomplete information. In *Proceedings of the 2013 International Conference on Autonomous Agents and Multi-agent Systems*, AAMAS '13, pages 715–722, Richland, SC, 2013. International Foundation for Autonomous Agents and Multiagent Systems

References

1. Baarslag T, Fujita K, Gerding EH, Hindriks KV, Ito T, Jennings NR, Jonker CM, Kraus S, Lin R, Robu V, Williams CR (2013) Evaluating practical negotiating agents: results and analysis of the 2011 international competition. Artif Intell 198:73–103
2. Baarslag T, Hindriks KV (2013) Accepting optimally in automated negotiation with incomplete information. In: Proceedings of the 2013 international conference on autonomous agents and multi-agent systems. AAMAS '13, Richland, pp 715–722. International Foundation for Autonomous Agents and Multiagent Systems
3. Ben Adar M, Sofy N, Elimelech A (2013). Gahboninho: strategy for balancing pressure and compromise in automated negotiation. In: Ito T, Zhang M, Robu V, Matsuo T (eds) Complex automated negotiations: theories, models, and software competitions, studies in computational intelligence, vol 435. Springer, pp 205–208
4. Bosse T, Jonker CM (2005) Human versus computer behaviour in multi-issue negotiation. In: Proceedings of the rational, robust, and secure negotiation mechanisms in multi-agent systems, RRS '05. IEEE Computer Society, Washington, pp 11–24
5. Coehoorn RM, Jennings NR (2004) Learning an opponent's preferences to make effective multi-issue negotiation trade-offs. In: Proceedings of the 6th international conference on electronic commerce. ICEC '04, ACM, New York, pp 59–68
6. DeGroot MH (1970) Optimal statistical decisions. McGraw-Hill, New York
7. Dirkzwager ASY, Hendrikx MJC, Ruiter JR (2013) The negotiator: a dynamic strategy for bilateral negotiations with time-based discounts. In: Ito T, Zhang M, Robu V, Matsuo T (eds) Complex automated negotiations: theories, models, and software competitions, Studies in computational intelligence, vol 435. Springer, pp 217–221
8. Faratin P, Sierra C, Jennings NR (1998) Negotiation decision functions for autonomous agents. Robot Auton Syst 24(3–4):159–182
9. Fatima SS, Wooldridge MJ, Jennings NR (2004) An agenda-based framework for multi-issue negotiation. Artif Intell 152(1):1–45

10. Ferguson TS (1989) Who solved the secretary problem? Stat Sci 4(3):282–289
11. Freeman PR (1983) The secretary problem and its extensions: a review. Int Stat Rev/Rev Int Stat 51(2):189–206
12. Frieder A, Miller G (2013) Value model agent: a novel preference profiler for negotiation with agents. In: Ito T, Zhang M, Robu V, Matsuo T (eds) Complex automated negotiations: theories, models, and software competitions, studies in computational intelligence, vol 435. Springer, pp 199–203
13. Kawaguchi S, Fujita K, Ito T (2013) AgentK2: compromising strategy based on estimated maximum utility for automated negotiating agents. In: Ito T, Zhang M, Robu V, Matsuo T (eds) Complex automated negotiations: theories. Models, and software competitions, studies in computational intelligence, vol 435. Springer, Berlin, pp 235–241
14. Kersten GE, Zhang G (2003) Mining inspire data for the determinants of successful internet negotiations. CEJOR 11(3):297–316
15. Kraus S, Wilkenfeld J, Zlotkin G (1995) Multiagent negotiation under time constraints. Artif Intell 75(2):297–345
16. Leonardz B (1973) To stop or not to stop. some elementary optimal stopping problems with economic interpretations. Almqvist and Wiksell, Stockholm
17. Lin R, Kraus S, Wilkenfeld J, Barry J (2008) Negotiating with bounded rational agents in environments with incomplete information using an automated agent. Artif Intell 172(6–7):823–851
18. Lin R, Kraus S, Baarslag T, Tykhonov D, Hindriks KV, Jonker CM (2014) Genius: an integrated environment for supporting the design of generic automated negotiators. Comput Intell 30(1):48–70
19. Mudgal C, Vassileva J (2000) Bilateral negotiation with incomplete and uncertain information: a decision-theoretic approach using a model of the opponent. In: Proceedings of the 4th international workshop on cooperative information agents IV, the future of information agents in cyberspace, CIA '00. Springer, London, pp 107–118
20. Rubinstein A (1982) Perfect equilibrium in a bargaining model. Econometrica 50(1):97–109
21. Stahl I (1972) Bargaining theory. Economic Research Institute, Stockholm
22. van Galen Last N (2012) Agent Smith: opponent model estimation in bilateral multi-issue negotiation. In: Ito T, Zhang M, Robu V, Fatima S, Matsuo T (eds) New trends in agent-based complex automated negotiations, studies in computational intelligence. Springer, Berlin, pp 167–174
23. van Krimpen T, Looije D, Hardheaded HS (2013) In: Ito T, Zhang M, Robu V, Matsuo T (eds) Complex automated negotiations: theories, models, and software competitions, studies in computational intelligence, vol 435. Springer, pp 223–227
24. Williams CR, Robu V, Gerding EH, Jennings NR (2011) Using gaussian processes to optimise concession in complex negotiations against unknown opponents. In: Proceedings of the twenty-second international joint conference on artificial intelligence, IJCAI'11, vol 1. AAAI Press, pp 432–438
25. Williams CR, Robu V, Gerding EH, Jennings NR (2012) Iamhaggler: a negotiation agent for complex environments. In: Ito T, Zhang M, Robu V, Fatima S, Matsuo T (eds) New trends in agent-based complex automated negotiations, studies in computational intelligence. Springer, Berlin, pp 151–158

Chapter 6
Measuring the Performance of Online Opponent Models

Abstract Decoupling agents as we have done in Chap. 3 helps us to focus on the individual components of a negotiating agent's design. One principal component of a negotiating agent's strategy is its ability to take the *opponent's preferences* into account. Every year, new negotiation agents are introduced with better learning techniques to model the opponent's preferences. Our main goal in this chapter is to *evaluate* and *compare* the performance of a selection of state-of-the-art online opponent modeling techniques in negotiation, and to determine under which circumstances they are beneficial in a real-time, online negotiation setting. Towards this end, we provide an overview of the factors influencing the quality of a model and we analyze how the performance of opponent models depends on the negotiation setting. This results in better insight into the performance of opponent models, and allows us to pinpoint a class of simple and surprisingly effective opponent modeling techniques that did not receive much previous attention in literature.

6.1 Introduction

Negotiating agents often keep their preferences private during the negotiation in order to avoid exploitation [1]; however, if an agent has no knowledge about the opponent's preferences, then this can result in a suboptimal outcome [2]. A common technique to counter this is *learning* the opponent's preference profile during the negotiation, which aids in increasing the quality of the negotiation outcome by identifying bids that are more likely to be accepted by the opponent [1–3].

If there have been previous negotiations with a similar opponent, the opponent model can be prepared *before* the start of the negotiation; we will refer to these models as *offline* models (for example [1]). Contrastingly, if the agent has to learn the preferences *during* the negotiation it performs *online* modeling (for example [2, 4, 5]).

We focus on *online* opponent models in a *single-shot* negotiation with *private* preference profiles; i.e., a setting in which an agent has no knowledge about the opponent's preference profile and no history of previous negotiations is available. There are opponent modeling techniques available for such settings, for example in

© Springer International Publishing Switzerland 2016
T. Baarslag, *Exploring the Strategy Space of Negotiating Agents*,
Springer Theses, DOI 10.1007/978-3-319-28243-5_6

the Automated Negotiating Agents Competition (ANAC) described in Appendix B. Despite ongoing research in this area, it is not yet clear how different approaches compare, and empirical evidence has raised the question whether using an opponent model is beneficial at all in such a setting. To illustrate: state-of-the-art agents, such as the top three agents of both ANAC 2010 (Appendix C) and ANAC 2011 (Appendix D) do not model the opponent, yet outperformed agents that do. One reason that opponent modeling does not guarantee a better outcome for an agent is that the model can be a poor representation of the opponent's preferences. If the model consistently suggests unattractive bids for the opponent, it may even be preferable to not employ one at all. Secondly, a time-based deadline introduces an additional challenge for online opponent modeling, as learning the model can be computationally expensive and can therefore influence the amount of bids that can be explored. More precisely, the gain in using the model should be higher than the loss in utility due to decreased exploration of the outcome space. We will refer to this as the *time/exploration trade-off*.

Apart from the inherent trade-off in opponent modeling, we are interested whether opponent models are accurate enough to provide gains at all, even when ignoring computational costs. To this end, we evaluate opponent models in two settings: a time-based and round-based negotiation protocol. This chapter compares a large set of opponent modeling techniques, which were isolated from state-of-the-art negotiation strategies. We measure their performance in various negotiation settings, and we provide a detailed overview of how the different factors influence the final negotiation outcome.

We introduce the negotiation setting and consider the difficulties in evaluating opponent models in Sect. 6.2. In Sect. 6.3 we introduce a method to quantify opponent model performance, after which we apply it to a set of models in Sect. 6.4. We formulate hypotheses and analyze the results in Sect. 6.5, and our chapter conclusions are outlined in Sect. 6.6.

6.2 Evaluating Opponent Models

We focus on a bilateral negotiations governed by the alternating-offers protocol, consistent with our definitions in Chap. 2. Recall from Sect. 2.2.3 that a negotiation scenario consists of the negotiation domain, which is *common knowledge*, together with a *privately-known* preference profile for each party. A preference profile is described by a utility function $u(x)$, which maps each possible outcome x in the negotiation domain to a utility in the range [0, 1]. We discuss opponent models that attempt to estimate the opponent's utility function $u'(x)$ during the negotiation.

The main goal of this chapter is to answer the following research question: "Under what circumstances is it beneficial to use an online opponent model in a real-time negotiation setting?". An answer to this question is not straightforward due to the time/exploration trade-off and potentially poor accuracy of a model. In particular, we want to answer the following:

1. Assuming *perfect* knowledge about the opponent's preferences, is there a significant performance gain in using this information compared with ignoring it?
2. Is there a significant performance gain from using an *online opponent model* in comparison to *not using a model*, assuming no prior knowledge is available?

The main difficulty in finding a conclusive answer to these questions, is that the performance of an opponent model depends on the negotiation setting. Therefore, we study an third, overarching research question:

3. How does the performance of using an opponent model depend on the negotiation setting?

6.2.1 *Influence of the Agent's Strategy*

Different agents apply their opponent model in different ways. There are two main factors in which the application of an opponent model by a bidding strategy can differ:

- *Type of information gained from the opponent model.* A bidding strategy can employ an opponent model for different reasons: for example, it can be employed to select the best bid for the opponent out of a set of similarly preferred bids [6, 7]; to select a bid that optimizes a weighted combination of both utility functions [4]; or to help estimate the utility of a specific outcome, such as the Nash-point [6].
- *Selecting a bid using an opponent model.* When a model is used to select a bid from a set of similarly preferred bids, the question still remains which one to choose. One can select the *best* bid for the opponent, but this may be suboptimal, as models may be inaccurate. An alternative is to select a bid from the set of n best bids [6].

Even when the factors above are taken into account, still care has to be taken to properly compare different models. We can only be fairly compare opponent models if the other components, such as bidding strategy and acceptance strategy are fixed.

6.2.2 *Influence of the Opponent's Strategy*

All opponent modeling techniques make certain assumptions about the opponent, so as to assign meaning to the observed behavior. If the opponent does not adhere to these assumptions, the model may not reflect reality well. The set of strategies against which a model is tested is a decisive factor when measuring its performance. Therefore, a set of opponents should contain both agents that fulfill the model's assumptions to determine its efficacy in optimal conditions; and agents that test the model's robustness by violating its assumptions.

The following assumptions were found by analyzing the models in Sect. 6.4.2:

1. *The concession of the opponent follows a particular function.* Some opponent modeling techniques assume that the opponent uses a given time-based bidding strategy. Modeling the opponent then reduces to estimating all issue weights such that the predicted utility by the modeled preference profile is close to the assumed utility. This assumption is violated when the utility of the opponent's bids strongly deviate from what is assumed.

2. *The first bid made by the opponent is the most preferred bid.* The best bid is the selection of the most preferred value for each issue, and thereby immediately reveals which values are the best for each issue. Many agents start with the best bid. This assumption is violated when the opponent's first bid is not the best; for example when an agent offers random bids with a utility above a constant.

3. *There is a direct relation between the preference of an issue and the times its value is significantly changed.* To learn the issue weights, some models assume that the amount of times the value of an issue is changed is an indicator for the importance of the issue. The validity of this assumption depends on the distribution of the issue and value weights of the opponent's preference profile and its bidding strategy.

4. *There is a direct relation between the preference of a value and the frequency it is offered.* A common assumption to learn the value weights is to assume that values that are more preferred are offered more often. Similar to the issue weights assumption, this assumption strongly depends on the agent's strategy and domain.

To summarize, to construct an opponent model a set of assumptions about the opponent's preference profile are required. The assumptions introduced above vary in validity and robustness depending on the opponent and negotiation scenario.

6.2.3 Influence of the Negotiation Scenario

We distinguish three main factors of a scenario that can influence the quality of an opponent model (see also Sect. 2.2.4):

1. *Domain size.* In general, the larger the domain, the less likely a bid is a Pareto-bid. Furthermore, domains with more bids are likely more computationally expensive to model. Therefore, the influence of the time/exploration trade-off is higher.

2. *Bid distribution.* The bid distribution quantifies how bids are distributed. We define bid distribution as the average distance of all bids to the nearest Pareto-bid. The bid distribution directly influences the performance gain attainable by a model.

3. *Opposition.* We define opposition as the distance from the Kalai-point to complete satisfaction $(1, 1)$. The opposition of a domain influences the number of possible agreements, and opponent models may be help in locating them more easily.

6.3 Measuring the Performance of Opponent Models

As we noted in the previous section, the effectiveness of an agent's opponent model is heavily influenced by the negotiation setting. In this chapter, we propose a careful measurement method of opponent modeling performance that can be interpreted as a first step towards creating a generic performance benchmark for the type of opponent models that we study here. The following sections discuss the four components of the method.

6.3.1 Negotiation Strategies of the Agents

We tested the performance of the opponent models by coupling them with a variety of negotiation strategies using our BOA architecture (Chap. 3). For the negotiation strategies of the agents in which the opponent models are embedded, we elected a variant of the standard time-dependent tactic [8] as defined in Sect. 2.3.3. This strategy is chosen for its simple behavior, which elicits regular behavior from its opponents; furthermore, adding a model may significantly increase its performance. Given a target utility, the adapted agent generates a set of similarly preferred bids and then selects one using the opponent model. We focus on selecting a bid from a set of similarly preferred bids, as this usage is commonly applied, for example in [7] and [5]. We embedded the models in four time-dependent agents ($e = 0.1; 0.2; 1.0; 2.0$). We opted for multiple agents as we believed that the concession speed can influence the performance gain.

The remaining issue in using an opponent model is which bid to select for the opponent given a set of similarly preferred bids. Given the approaches in Sect. 6.2.1, we opted to have the models select the best bid for the opponent, as this approach is most differentiating: it leads to better performance of the more accurate opponent models.

6.3.2 Negotiation Strategies of the Opponents

This section discusses the opponents selected using the guidelines outlined in Sect. 6.2.2. The set of opponent strategies consists of three cooperative agents, which should be easy to model as their concession speed is high, and five competitive agents. The set of conceding agents consists of two *time-dependent agents* with high concession speeds $e \in \{1, 2\}$, and the *Offer Decreasing* agent, which offers the set of all possible bids in decreasing order of utility. The set of competitive agents contains two *time-dependent agents* with low concession speeds $e \in \{0.0, 0.2\}$, and the ANAC agents *Gahboninho*, *HardHeaded*, and *IAMcrazyHaggler*.

Given the five opponent modeling assumptions introduced in Sect. 6.2.2, the first assumption about the opponent's decision function fails in general, as an opponent in practice never completely adheres to the assumed decision function. The second assumption holds for all agents except *IAMcrazyHaggler*, whose first bid is randomly picked. The other three assumptions are typical for the frequency models. It is not possible to adhere to or violate these assumptions completely, as they depend both on the negotiation scenario structure and opponents

6.3.3   Negotiation Scenarios

As we stated in Sect. 6.2.3, the *domain size*, *bid distribution*, and *opposition* of a negotiation scenario are all expected to influence an opponent model's performance, and therefore we aimed for a large spread of the characteristics of the scenarios. In total seven negotiation scenarios were selected, as depicted in Table 6.1. Full details on the scenarios can be found in the references provided in Table 6.1, and in Appendix C and D.

6.3.4   Quality Measures for Opponent Models

The quality of an opponent model can be measured in two ways: a black box approach, in which *performance measures* evaluate the quality of the outcome; and a white box view, which uses *accuracy measures* capable of considering the internal design of a strategy and revealing the accuracy of the estimation of the opponent's preferences.

The work in this chapter focuses on the performance measures shown in Table 6.2 (for more information on performance measures we refer to Sect. 2.4.3). We will compare models using a white box approach in Chap. 7.

We employ six different performance measures. First, we keep track of the *utility performance* of an individual opponent model, which is measured as the average score of the agents employing it against the selected opponents on all negotiation

Table 6.1 Characteristics of the negotiation scenarios

Scenario name	Size	Bid distrib.	Opposition
Car [9]	15625 (*med.*)	0.136 (*low*)	0.095 (*low*)
Grocery [9]	1600 (*med.*)	0.492 (*high*)	0.191 (*med.*)
Company Acquisition [9]	384 (*low*)	0.121 (*low*)	0.125 (*low*)
Itex versus Cypress [10]	180 (*low*)	0.222 (*med.*)	0.431 (*high*)
Laptop [9]	27 (*low*)	0.295 (*med.*)	0.178 (*med.*)
Employment contract [11]	3125 (*med.*)	0.267 (*med.*)	0.325 (*high*)
Travel [12]	188160 (*high*)	0.416 (*high*)	0.230 (*med.*)

Table 6.2 Overview of the performance measures

Performance measure	Description
Average utility [2, 9, 13]	Average score of the agents against selected opponents on all negotiation scenarios
Average time of agreement [14]	Average time required to reach an agreement
Average rounds [3, 13]	Average rounds a negotiation lasts. In a rounds-based setting, less means more accurate
Average Pareto distance of agreement [9, 15]	Average minimal distance to the Pareto-frontier. Lower is better
Average Kalai distance of agreement [15]	Average distance to the Kalai-point. Lower means more fair
Average Nash distance of agreement [15]	Average distance to the Nash-point. Lower means more fair

scenarios. We also measure the *average time of agreement* and the *average amount of rounds* that the negotiation takes. Finally, we test the *average distance from the outcome to the Pareto-frontier, Kalai-point, and Nash-point* of all negotiations that result in an agreement.

6.4 Experiments

We applied the method described in the previous section to our experimental setup below in order to answer the research questions introduced in Sect. 6.2.

6.4.1 Experimental Setup

To analyze the performance of different opponent models, we employed GENIUS to evaluate the automated negotiators' strategies and their components. The experiments are subdivided into two categories: we use a standard *time-based protocol*, as well as a *round-based protocol*. In total, we ran 17920 matches, which on a single computer takes nearly 2 months.

Our main interest goes out to the real-time setting, as this protocol features the time/exploration trade-off. We applied our benchmark to the set of models using the time-based protocol. Each match features a real-time deadline set at three minutes. The full tournament consists of 17920 matches, which takes more than an month to run on a single computer.

In the round-based protocol the same approach is applied, but in this case, time does not pass within a round, giving the agent infinite time to update its model. This provides valuable insights into the best *theoretical* result an opponent model can achieve.

6.4.2 Opponent Models

We compare the performance of the opponent models used in ANAC (Appendix B), which is a yearly international competition in which negotiating agents compete on multiple domains. Each year, the competition leads to the introduction of new negotiation strategies with novel opponent models, as the utility function of each player is private information and hence has to be learned. The utility functions of the agents are *linearly additive*; that is, the overall utility consists of a weighted sum of the utility for each individual issue. The setting of ANAC is consistent with the preliminaries in this chapter; i.e., the negotiation setting consists of an alternating offers protocol with discrete negotiation scenarios and real-time deadlines.

We specifically opted to use agents that participated in ANAC for the following reasons: the agents are designed for one consistent negotiation setting, which makes it possible to compare them fairly; their implementation is publicly available; and finally, we believe that the agents and opponent models represent the current state-of-the-art. We used modeling techniques from ANAC 2010, ANAC 2011, and a selection of opponent models designed for ANAC 2012.

We isolated the opponent models from the agents and reimplemented them as separate generic components to be compatible with all other agents using our BOA framework (Chap. 3). As discussed in Sect. 6.2.1, this setup allows us to equip a single negotiation strategy with various opponent models, which makes it straightforward to fairly compare the different modeling techniques.

Table 6.3 provides a summary of the online opponent models used in our experiments, with references to the work in which they are described. We did not include the *Bayesian Model* from [2] and the *FSEGA Bayesian Model* [16], even though they fitted our setup, as both models were not designed to handle domains containing more than a 1000 bids. We are aware that many alternative opponent modeling techniques exist [2, 3, 17, 18]; however, for our negotiation setting, this was the largest set available of comparable opponent modeling techniques.

Based on our analysis, we found that in our selection two approaches to opponent modeling are prominent: *Bayesian opponent models* and *Frequency models*.

Bayesian opponent models generate hypotheses about the opponent's preferences [2]. The models presuppose that the opponent's strategy adheres to a specific decision function; for example a time-dependent strategy with a linear concession speed. This is then used to update the hypotheses using Bayesian learning.

Frequency models learn the issue and value weights separately. The issue weights are usually calculated based on the frequency that an issue *changes* between two offers. The value weights are oten calculated based on the frequency of *appearance* in offers.

Both modeling approaches are prone to failure as they rely on a subset of the assumptions introduced in Sect. 6.2.2. More specifically, Bayesian models make strong assumptions about the opponent's strategy, whereas frequency models assume knowledge about the value distribution of the issues of a preference profile and place

Table 6.3 Overview of the online opponent models and their modeling assumptions (M)

Model	Description	M
No Model	No knowledge about the preference profile	–
Perfect Model	Perfect knowledge about the preference profile	–
Bayesian Scalable Model [2]	This model learns the issue and value weights separately using Bayesian learning. Each round, the hypotheses about the preference profile are updated based assuming that the opponent conceded a constant amount	1
IAMhaggler Bay. Model [7]	Efficient implementation of the *Bayesian Scalable Model* in which the opponent is assumed to use a particular time-dependent decision function	1
HardHeaded Freq. Model [5]	This model learns the issue weights based on how often the value of an issue changes between turns. The value weights are determined based on the frequency in which they have been offered	3, 4
Smith Freq. Model [4]	Similar to the *HardHeaded Frequency Model*, but less efficient. The issue weights depends on the relative frequency of the most offered values	3, 5
Agent X Freq. Model	This model is a more complex variant of the *HardHeaded Frequency Model* that also takes the opponent's tendency to repeat bids into account	3, 4
N.A.S.H. Freq. Model	In contrast to *HardHeaded Frequency Model*, this model learns the issue weights based on the frequency that the assumed best value is offered	2, 4

weak restrictions on the opponent's negotiation strategy. Generally, the Bayesian models are far more computationally expensive; however, it is unknown if they are more accurate.

6.5 Results

Below we analyze the outcomes of the experiment to provide an answer to the research questions in the form of Hypotheses 6.1 through 6.6. We first discuss the overall gain in performance when using perfect knowledge versus online opponent modeling. Section 6.5.2 provides an answer to the final research question on how the negotiation setting influences the performance of an opponent model.

6.5.1 Overall Performance of Opponent Models

Our experimental results for a selection of the quality measures described in Sect. 6.3.4 are shown in Table 6.4 for both the time-based and round-based protocol.

Table 6.4 Performance of all models on a set of quality measures for both protocols

Quality measures	Perfect	HH. FM	Ag.X FM	Nash FM	IAH. BM	Smith FM	None	Scal. BM
Time-based								
Avg. utility	0.7285	0.7260	0.7257	0.7257	0.7178	0.7156	0.7125	**0.7077**
Avg. time of agr.	**0.4834**	0.4865	0.4867	0.4865	0.4958	0.4937	0.5022	0.5055
Avg. rounds	7220	7218	7231	7198	7004	**4745**	7352	4836
Avg. Pareto dist. of agr.	**0.0007**	0.0017	0.0015	0.0018	0.0069	0.0068	0.0059	0.0071
Avg. Kalai dist. of agr.	**0.2408**	0.2434	0.2447	0.2428	0.2515	0.2474	0.2683	0.2561
Avg. Nash dist. of agr.	**0.2442**	0.2471	0.2481	0.2483	0.2541	0.2500	0.2721	0.2594
Rounds-based								
Avg. utility	0.7235	0.7196	0.7191	0.7192	0.7111	0.7199	**0.7050**	0.7124
Avg. time of agr.	**0.4928**	0.4975	0.4978	0.4977	0.5058	0.4974	0.5136	0.5038
Avg. rounds	**2508**	2531	2533	2533	2572	2531	2567	2562
Avg. Pareto dist. of agr.	**0.0010**	0.0029	0.0023	0.0028	0.0073	0.0026	0.0066	0.0063
Avg. Kalai dist. of agr.	**0.2332**	0.2380	0.2395	0.2380	0.2456	0.2369	0.2614	0.2445
Avg. Nash dist. of agr.	**0.2370**	0.2403	0.2437	0.2404	0.2516	0.2403	0.2644	0.2472

A bold item indicates the lowest value for a quality measure, a underlined item the highest

Before we analyze the performance gain of online opponent models, we first answer the question whether perfect knowledge aids in improving the negotiation outcome at all:

Hypothesis 6.1 Usage of the perfect model by a negotiation strategy leads to a significant performance gain in comparison to not using an opponent model.

We expected that perfect knowledge about the opponent's preferences would significantly improve performance of an agent. Our main aim here was not to reconfirm the already widely acknowledged benefits of integrative bargaining, but to analyze whether our experimental setup is a valid instrument for measuring the learning effect in other types of settings. Our expectation is confirmed by the experiment, as the *Perfect Model* yields a significant performance increase on all quality measures (except average rounds) for both protocols. For the real-time protocol, the difference between the best online opponent model (*HardHeaded Frequency Model*) and *No Model* is 0.0135; for the round-based protocol it is 0.0144 (*Smith Frequency Model*). Note that while the gains are small, there are three small domains where opponent modeling does not result in significant gains. If we solely focus on the large *Travel* negotiation scenario, then the gain relative to *No Model* becomes 0.0413 for the *Perfect Model*. Especially note the improvement in distance between the outcome and Pareto-frontier, and the earlier agreements, in Table 6.4. This leads us to conclude that using an opponent model leads to better performance as it aids in increasing the quality of the outcome.

> **Hypothesis 6.2** Usage of an online opponent model leads to a significant performance gain when time is not an issue. Online opponent modeling does not yield the same benefit in a real-time setting because of the time/exploration trade-off.

We noted previously that in some cases, ANAC agents that do not model the opponent can outperform agents that do, and such agents have even won the competition (see Tables C.2 and D.2). This led us to believe that online modeling does not benefit the agents, either because it misrepresents the preferences, or by taking too much time in a time-sensitive setting.

This is why it came as a surprise that in *both* the time- and round-based protocol, online opponent models performed significantly better on all quality measures. For the time-based protocol the best online opponent models are the frequency models, except for the *Smith Frequency Model* who scores badly in this case. However, for the round-based protocol, the *Smith Frequency Model* is actually best. This is caused by the time/exploration trade-off, because the model is computationally expensive as indicated by the small amount of bids offered in the time-based protocol.

Surprisingly the worst performance on a quality measure is not always made by using *No Model*. For example in the time-based experiment the *Bayesian Scalable Model* has the worst performance. The Bayesian model of *IAMhaggler* however, performs much better, but disappoints in the round-based protocol. We believe this can be attributed to its updating mechanism: only unique bids are used to update the model, which speeds-up updating but can result in poor performance against slowly conceding agents that offer the same bid multiple times.

In conclusion, online opponent model can result in significant gains and surprisingly, frequency models lead to the largest gains, outperforming the Bayesian models. We believe that the winners of ANAC could have performed even better by learning the opponent's preferences with a frequency model. The success of the frequency model can be attributed to its simplicity and hence faster performance, and to the fact that it is more robust by making weaker assumptions about the strategy of the opponent in comparison to the Bayesian modeling approaches.

6.5.2 Influence of the Negotiation Setting

We will now discuss the influence of each of the three components of the negotiation setting on the quality of an opponent model, following the structure of Sect. 6.2.

Table 6.5 Utility of each opponent model relative to using *No Model* for each agent

Opponent model	Agents			
	$e = 0.1$	$e = 0.2$	$e = 1$	$e = 2$
Perfect Model	0.0180	0.0164	0.0152	0.0144
HardHeaded Freq. Model	0.0156	0.0137	0.0118	0.0128
Agent X Freq. Model	0.0161	0.0137	0.0116	0.0113
N.A.S.H. Freq. Model	0.0166	0.0129	0.0108	0.0121
IAMhaggler Bay. Model	0.0084	0.0055	0.0033	0.0039
Smith Freq. Model	−0.0031	0.0020	0.0071	0.0063
Bayesian Scalable Model	−0.0050	−0.0058	−0.0032	−0.0053

6.5.3   Influence of the Agent's Strategy

The performance gain of using an opponent model necessarily depends on the strategy in which it is embedded (cf. Chap. 3). Table 6.5 provides an overview of the relative gain in comparison to *No Model* for all opponent models in the time-based experiment. Based on the results, we have tested the following hypothesis:

Hypothesis 6.3 The more competitive an agent, the more it benefits from using an opponent model.

At each turn of a negotiation session, a set of possible agreements can be defined. This is the intersection of two sets: the set of bids that an agent considers for offering, and the set of all bids acceptable to the opponent. The more competitive the agent, the smaller the intersection between the two sets. When an agent concedes, the number of possible agreements increases at the cost of utility. An opponent model can help in finding possible agreements, preventing concession and therefore loss in utility. We therefore expected the gain for competitive agents to be higher, as the set of possible agreements each turn is smaller, and therefore an optimal bid is more easily missed by an agent not employing an opponent model. This is especially decisive in the last few seconds of the negotiation, when many agents concede rapidly to avoid non-agreement.

The hypothesis is confirmed by our experiments. In Table 6.5 there is a negative correlation between the concession speed and relative gain in performance. If we ignore the results of the three worst performing models, a small—albeit statistically significant—negative correlation of −0.508 is found.

Table 6.6 Utility of each opponent model relative to using *No Model* for each opponent

Opponent model	Opponent agents							
	TDT 0.0	TDT 0.2	TDT 1.0	TDT 2.0	OD	Gah.	HH.	IcH.
Perfect	0.0085	0.0015	0.0008	0.0022	0.0060	0.0676	0.0015	0.0399
HH. Freq. Model	0.0085	0.0013	−0.0002	0.0019	0.0060	0.0515	0.0000	0.0388
Agent X Freq. Model	0.0085	0.0019	0.0002	−0.0036	0.0058	0.0561	0.0009	0.0285
N.A.S.H. Freq. Model	0.0085	0.0005	−0.0005	0.0020	0.0065	0.0507	0.0037	0.0336
IAH. Bay. Model	0.0000	0.0003	−0.0021	−0.0001	−0.0046	0.0511	0.0039	**−0.0066**
Smith frequency	**−0.0038**	−0.0023	−0.0019	0.0007	−0.0113	**0.0357**	**−0.0224**	0.0297
Bay. Scalable Model	0.0000	**−0.0033**	**−0.0055**	**−0.0058**	**−0.0535**	0.0458	−0.0128	−0.0036

A bold item indicates the lowest value for a domain, a underlined item the highest

6.5.4 Influence of the Opponent's Strategy

The opponent's behavior also has an important impact on the performance of an opponent model. Based on the results shown in Table 6.6, we test the three hypotheses below.

> **Hypothesis 6.4** An agent benefits more from an opponent model against competitive agents.

Intuitively, the more competitive the opponent, the more useful the opponent model as the set of possible agreements is smaller, analogous to Hypothesis 6.3. Therefore, we expected the highest gain against the competitive agents *Gahboninho*, *HardHeaded*, and *IAMcrazyHaggler*. However, in Table 6.6 only the gain for *Gahboninho* and *IAMcrazyHaggler* is very high.

For *HardHeaded*, we believe this can be attributed to the agent using an opponent model itself. If the opponent uses a well-performing opponent model, then the performance gain of an opponent model can be expected to be lower, as the opponent is already able to make Pareto-optimal bids. Our experiment appears to the confirm this hypothesis in the case of playing against *HardHeaded*, whose well-performing opponent model seems to diminish the effect of opponent modeling by the other side.

Concluding, given the results of our experiment, we believe that the hypothesis holds, at least for consistently competitive opponents without an opponent model.

Hypothesis 6.5 Frequency models are more robust against opponents employing a random tactic than the Bayesian models.

In order to estimate the opponent's utility of a certain bid, both types of models make certain assumptions about the opponent. The Bayesian opponent models assume that the opponent follows a particular decision function through time (cf. modeling assumption 1 in Sect. 6.2.2), while the frequency models assume higher valued bids are offered more often (cf. modeling assumptions 3 and 4). Many opponent strategies do not adhere to these assumptions, which causes the learning models to make wrong predictions when playing against them. For example, opponents such as *IAMcrazyHaggler* who employ a random negotiation strategy, explicitly violate the assumptions of both models. For the Bayesian learning models, this means the opponent preferences will be estimated incorrectly, and more so through time. The frequency models however, are much more robust, not only in the sense that a negotiation tactic has a greater chance to satisfy its assumptions, but more significantly: it is less sensitive to a tactic violating its assumptions. For instance, in the case of *IAMcrazyHaggler*, it will deduce that it equally prefers any bid it has offered so far—which, in this case, is exactly right.

We therefore expected relatively poor performance from the Bayesian models. This hypothesis is confirmed by our experiment: the frequency models have a high performance gain against *IAMcrazyHaggler*, whereas using the Bayesian models is even worse than not using an opponent model at all.

6.5.5   *Influence of the Negotiation Scenario*

The performance of an opponent model is influenced by the characteristics of the negotiation scenario, such as amount of bids, distribution of the bids, and the opposition of the domain. Table 6.7 provides an overview of the relative gain of all opponent models in comparison to *No Model* for in the time-based experiment. Based on these results, we formulate Hypothesis 6.6.

Hypothesis 6.6 The higher the amount of bids, bid distribution, or opposition of a scenario, the more an agent benefits from using an opponent model.

We anticipated the bid distribution to be the major factor determining the performance gain of an opponent model. If the bid distribution is high, then the Pareto-frontier is more sparse. This means a higher gain can be expected of utilizing an opponent model to locate bids close to the Pareto-frontier. This is confirmed by our experiments, as we found a strong Pearson correlation of 0.778 between the bid

Table 6.7 Gain of each model relative to using *No Model* for each scenario parameter

Parameter	Model	Parameter value		
		Low	Medium	High
Size	Perfect	0.001	0.022	0.041
	Best 4	0.002	0.018	0.039
Bid distribution	Perfect	0.001	0.013	0.035
	Best 4	−0.001	0.010	0.034
Opposition	Perfect	0.001	0.023	0.020
	Best 4	−0.001	0.022	0.016

distribution and the performance gain of the best four models, and 0.701 if we solely focus on the perfect opponent model. Therefore we confirm this sub-hypothesis.

Another factor is the size of the negotiation domain. If a domain contains more bids, then there are relatively less bids that are Pareto-optimal, so an opponent model can aid more in identifying them. On the other hand, opponent models are more computationally expensive on the larger domains. Despite this effect, we found a strong Pearson correlation between the amount of bids and the performance gain: 0.631 for the best four models, and 0.596 when using the perfect model.

The final factor is the opposition of the scenario. Intuitively, if the opposition is higher, then there are less possible agreements. Opponent models can aid in identifying these rare acceptable bids, thereby preventing break-offs and unnecessary concessions. Nevertheless, if the opposition is high, then the bids are also relatively closer to the Pareto-optimal frontier, which renders it more difficult for an opponent model to make a significant impact on the negotiation outcome. Despite this effect, we expected that higher opposition would lead to higher performance gain. However, in our experiments we noted only a small positive Pearson correlation of 0.256 for the best four models and 0.262 for the perfect model. Based on these results we are unable to draw a conclusion, which leads us to believe the two mentioned effects cancel each other out, making the other two characteristics of the scenario decisive in the effectiveness of a model.

6.6 Conclusion

This chapter evaluates and compares the performance of a selection of state-of-the-art online opponent models. Our main goal was to evaluate if, and under what circumstances, opponent modeling is beneficial.

Measuring the performance of an opponent model is not trivial, as the details of the negotiation setting affects the effectiveness of the model. Furthermore, while we know an opponent model improves the negotiation outcome in general, the role of time should be taken into account when considering *online* opponent modeling in a real-time negotiation because of the time/exploration trade-off: a computationally expensive model may produce predictions of better quality, but in a real-time setting it may lead to less bids being explored, which may harm the outcome of the negotiation.

Based on an analysis of the contributing factors to the quality of an opponent model, we formulated a measurement method to quantify the performance of online opponent models and applied it to a large set of state-of-the-art opponent models. We analyzed two main types of opponent models: frequency models and Bayesian models. We noted that the time/exploration trade-off is indeed an important factor to consider in opponent model design of both types. However, we found that the best performing models did not suffer from the trade-off, and that most—but not all—online opponent models result in a significant improvement in performance compared with not using a model; not only because the deals are made faster, but also because the outcomes are on average significantly closer to the Pareto-frontier. A main conclusion of our work is that we noted that frequency models consistently outperform Bayesian models. This is not only because they are faster, because the effect remains in a round-based setting. This suggests that frequency models combine the best of both worlds. Surprisingly, despite their performance, frequency models have not received much attention in literature.

Our other main conclusion concerns the effects of the negotiation setting on an opponent model's effectiveness. We found that the more competitive an agent, or its opponent, the more benefit an opponent model provides. In addition, we found that the higher the size or the bid distribution of a scenario, the higher the gain of using a model. In the next chapter, we investigate this more thoroughly, by studying the interaction between the performance of an opponent model and its accuracy through time.

This chapter is based on the following publications: [19]

Tim Baarslag, Mark J.C. Hendrikx, Koen V. Hindriks, and Catholijn M. Jonker. Measuring the performance of online opponent models in automated bilateral negotiation. In Michael Thielscher and Dongmo Zhang, editors, *AI 2012: Advances in Artificial Intelligence*, volume 7691 of *Lecture Notes in Computer Science*, pages 1–14. Springer Berlin Heidelberg, 2012

References

1. Coehoorn RM, Jennings NR (2004) Learning an opponent's preferences to make effective multi-issue negotiation trade-offs. In: Proceedings of the 6th international conference on Electronic commerce, ICEC '04. ACM, New York, pp 59–68
2. Hindriks KV, Tykhonov D (2008) Opponent modelling in automated multi-issue negotiation using bayesian learning. In: Proceedings of the 7th international joint conference on Autonomous agents and multiagent systems, AAMAS '08. International foundation for autonomous agents and multiagent systems, vol 1. Richland, SC, pp 331–338
3. Dajun Z, Sycara KP (1998) Bayesian learning in negotiation. Int J Hum Comput Stud 48(1):125–141

4. van Galen Last N (2012) Agent Smith: opponent model estimation in bilateral multi-issue negotiation. In: Ito T, Zhang M, Robu V, Fatima S, Matsuo T (eds) New trends in agent-based complex automated negotiations. Studies in computational intelligence. Springer, Berlin, pp 167–174
5. van Krimpen T, Looije D, Hajizadeh S (2013) Hardheaded. In: Ito T, Zhang M, Robu V, Matsuo T (eds) Complex automated negotiations: theories, models, and software competitions. Studies in computational intelligence, vol 435. Springer, Berlin, pp 223–227
6. Baarslag T, Hindriks KV, Jonker CM (2013) A tit for tat negotiation strategy for real-time bilateral negotiations. In: Ito T, Zhang M, Robu V, Matsuo T (eds) Complex automated negotiations: theories, models, and software competitions. Studies in computational intelligence, vol 435. Springer, Berlin, pp 229–233
7. Williams CR, Robu V, Gerding EH, Jennings NR (2012) Iamhaggler: a negotiation agent for complex environments. In: Ito T, Zhang M, Robu V, Fatima S, Matsuo T (eds) New trends in agent-based complex automated negotiations. Studies in Computational Intelligence. Springer, Berlin, pp 151–158
8. Faratin P, Sierra C, Jennings NR (1998) Negotiation decision functions for autonomous agents. Robot Auton Syst 24(3–4):159–182
9. Baarslag T, Katsuhide F, Gerding Enrico H, Hindriks Koen V, Takayuki I, Jennings Nicholas R, Jonker Catholijn M, Sarit K, Raz L, Valentin R, Williams Colin R (2013) Evaluating practical negotiating agents: results and analysis of the 2011 international competition. Artif Intell 198:73–103
10. Kersten Gregory E, Grant Z (2003) Mining inspire data for the determinants of successful internet negotiations. Central Eur J Oper Res 11(3):297–316
11. Leigh T (2000) The Mind and heart of the negotiator, 3rd edn. Prentice Hall Press, Upper Saddle River
12. Baarslag T, Hindriks KV, Jonker CM, Kraus S, Lin R (2012) The first automated negotiating agents competition (ANAC 2010). In: Ito T, Zhang M, Robu V, Fatima S, Matsuo T (eds) New trends in agent-based complex automated negotiations. Studies in computational intelligence, vol 383. Springer, Berlin, pp 113–135
13. Klos Tomas B, Koye S, La Poutré JA (2011) Automated interactive sales processes. IEEE Intell Syst 26(4):54–61
14. Baarslag T, Hindriks KV, Hendrikx MJC, Dirkzwager ASY, Jonker CM (2012) Decoupling negotiating agents to explore the space of negotiation strategies. In: Proceedings of the fifth international workshop on agent-based complex automated negotiations (ACAN 2012)
15. Hindriks KV, Tykhonov D (2010) Towards a quality assessment method for learning preference profiles in negotiation. In: Ketter W, La Poutré Johannes A, Sadeh N, Shehory O, Walsh W (eds) Agent-mediated electronic commerce and trading agent design and analysis. Lecture notes in business information processing, vol 44. Springer, Berlin, pp 46–59
16. Şerban LD, Silaghi GC, Litan CM (2012) AgentFSEGA—time constrained reasoning model for bilateral multi-issue negotiations. In: Ito T, Zhang M, Robu V, Fatima S, Matsuo T (eds) New trends in agent-based complex automated negotiations. Series of studies in computational intelligence. Springer, Berlin, pp 159–165
17. Buffett S, Spencer B (2007) A bayesian classifier for learning opponents' preferences in multi-object automated negotiation. Electron Commer Res Appl 6(3):274–284
18. Lin R, Kraus S, Wilkenfeld J, Barry J (2008) Negotiating with bounded rational agents in environments with incomplete information using an automated agent. Artif Intell 172(6–7):823–851
19. Baarslag T, Hendrikx MJC, Hindriks KV, Jonker CM (2012) Measuring the performance of online opponent models in automated bilateral negotiation. In: Thielscher M, Zhang D (eds) AI 2012: Advances in artificial intelligence, vol 7691., Lecture notes in computer science. Springer, Berlin, Heidelberg, pp 1–14

Chapter 7
Predicting the Performance of Opponent Models

Abstract The quality of an opponent model can be measured in two different ways. One, which we extensively covered in Chap. 6, is to use the agent's *performance* as a benchmark for the model's quality. The other is to directly evaluate its *accuracy* by using *similarity measures*. Both methods have been used extensively, and both have their distinct advantages and drawbacks. Our work in this chapter bridges the gap between the two approaches by investigating a large set of opponent modeling techniques in different negotiation settings, measuring both their accuracy through time and their performance. We review all ways to measure the accuracy of an opponent model and we then analyze how changes in accuracy translate into performance differences. Moreover, we pinpoint the best *predictors* for good performance. This leads us to new insights in how to construct an opponent model, and what we need to measure when optimizing performance.

7.1 Introduction

A major challenge in automated negotiation is that agents usually keep their preference information private to avoid exploitation [1, 2]. When the agents have limited knowledge of the other's preferences, the agents may fail to reach an optimal outcome as they cannot take the opponent's desires into account [3].

In order to improve the efficiency of the negotiation and the quality of the outcome, agents may construct a *model* of the opponent's preferences, which aids them in estimating the information that is kept private [1–3]. Over time, a large number of such opponent models have been introduced, based on different learning techniques and underlying assumptions, and multiple methods have been used to compare their quality. The different evaluation methods for opponent models make it hard to compare different approaches, as each method has its unique scope of application, together with different advantages and drawbacks. From an engineering perspective, it still remains unclear which opponent model to choose in a particular negotiation setting.

As we outlined in Sects. 2.4.3 and 2.4.4, there are two popular ways to measure the quality of an opponent model:

© Springer International Publishing Switzerland 2016
T. Baarslag, *Exploring the Strategy Space of Negotiating Agents*,
Springer Theses, DOI 10.1007/978-3-319-28243-5_7

1. *Performance measures* evaluate the quality of the outcome, usually measured in utility gain, or distance of the agreement to the Pareto frontier. With this method, the success of an opponent model is expressed in terms of the negotiation *result* (as opposed to the whole negotiation *process*).
2. *Accuracy measures* aim to determine the quality of a model in a more fundamental way, by quantifying how well the opponent model *represents* the real preferences of the opponent, using a certain similarity measure. An example is the correlation between the estimated and the real outcome space, or the percentage of correctly inferred Pareto optimal outcomes.

There are various authors that evaluate their opponent model with *performance measures* (e.g. [1–3]). Using a performance measure has one very important quality: it measures exactly what needs to optimized, namely the net effect an opponent model has on the negotiation result. On the other hand, because performance measures are only able to demonstrate improvement of the end result, they may not provide insight into why or how an opponent model works; that is, they measure the result obtained by the negotiation agent as a whole, of which the opponent model is only a single component. This makes the performance measure very sensitive to the specifics of the experimental setup. Moreover, there is usually no clear upper bound in performance gain, so it remains unclear what the highest attainable result is.

Other authors prefer to use *accuracy measures* to evaluate their model (for example [4–7]). The main advantage of this approach is that it directly assesses the quality of a model, independent of other factors such as bidding strategy or acceptance strategy. Secondly, it is easier to compare accuracy results between different experimental setups, and to track the accuracy of a model over the course of the negotiation. This, in turn, can reveal valuable information about the reasons for a model's success.

There are also drawbacks of using accuracy measures in negotiation, two of which we will address in this chapter. First, it is currently unclear what effect a more accurate opponent model has on the negotiation outcome. It could very well be that from some point, increased accuracy does not translate into better performance. An 80 % accurate model for example, could perform just as well as a perfect model. Second, there are many accuracy measures to choose from, and it is currently unknown which accuracy measure should be selected to ensure a good overall end result; that is, we would like to know what accuracy measure best *predicts* an improvement in performance.

Our work in this chapter bridges the gap between both approaches by considering opponent models from both a performance and an accuracy perspective. We first test many current opponent modeling techniques in different negotiation settings, measuring both their accuracy through time and their performance. We then analyze how changes in accuracy translate into performance differences. Moreover, we review all ways to measure the accuracy of an opponent model, and we pinpoint the best *predictors* for good performance.

The remainder of this chapter is organized as follows. In Sect. 7.2 the setting and terminology used in this chapter and introduces our research questions and experiments, followed by a discussion of the results in Sect. 7.3. Finally, Sect. 7.4 outlines the conclusions of this chapter.

7.2 Measuring the Quality of Opponent Models

The aim of this chapter is to answer three research questions:

1. How does the accuracy of opponent models depend on negotiation factors, such as domain size, or time?
2. What is the relationship between the accuracy of an opponent model and its expected performance gain?
3. What accuracy measures are the best predictors for performance gain?

To answer our three questions, we first outline our selection of opponent models (Sect. 7.2.2) and the accuracy measures incorporated in our method (Sect. 7.2.3). Next, we discuss the experimental setup (Sects. 7.2.4 and 7.2.5). We start with the preliminaries of this chapter.

7.2.1 Preliminaries

We use the same negotiation setting as in Chap. 6; however, instead of a continuous time line, we use a *discrete* time line here, since we want to sample not only the performance, but also the quality of opponent models at regular time intervals. This means a deadline occurs after a specified number of rounds N, and both agents receive utility 0 if they do not succeed in reaching an agreement before this time.

Recall from Sect. 2.2.3 that the negotiation domain, which specifies all possible bids, is *common knowledge*, while the preferences for each party is *private information*. We discuss opponent models that attempt to estimate the opponent's utility function $u_{op}(\omega)$ while relying solely on the information gathered during the negotiation. We restrict ourselves to linear additive preference profiles in this work, as there exists no large set of comparable models for non-linear preferences. The utility $u(\omega)$ of an outcome $\omega \in \Omega$ is therefore assumed to be computed as a weighted sum (as specified by the *issue weights* w_i) of *value weights* $e_i(\omega_i)$:

$$u(\omega) = \sum_{i=1}^{n} w_i \cdot e_i(\omega_i). \tag{7.1}$$

7.2.2 Selection of Opponent Models

We compare a large set of state of the art opponent models, which were applied in the Automated Negotiating Agents Competition (ANAC, Appendix B). ANAC is a yearly international competition in which negotiation agents compete on a set of scenarios that are unknown beforehand. As in Chap. 6, our reason for including this set of models is threefold: first, they represent the state of the art; second, to our knowledge, they are the largest set of techniques designed for one common setting consistent with ours; and finally, their code is publicly available. Table 7.1 gives an overview of all models we evaluated, including three theoretical baselines. We distinguish four types of opponent models:

Table 7.1 Overview of opponent models

Bayesian Models	
Bayesian Scalable Model [3]	Estimates the issue and value weights separately, using Bayesian learning. The opponent is assumed to concede a constant amount per round
IAMhaggler Bayesian Model [8]	A Bayesian model in which the opponent is assumed to use a particular time-dependent strategy and only unique bids are used to update the model
Frequency Models	
HardHeaded Frequency Model [9]	Learns the issue weights based on how often the value of an issue changes. The value weights are estimated based on the frequency they are offered
Smith Frequency Model [10]	Learns the value weights based on frequency they are offered. The issue weights are estimated based on the distribution of the values
Agent X Frequency Model	A variant of the HardHeaded Frequency Model that takes the opponent's tendency to repeat bids into account
N.A.S.H. Frequency Model	Learns the issue weights based on how often the best value for each issue is offered. The value weights are estimated based on their frequency
Value Models	
AgentLG Value Model	Estimates the value weights based on the frequency they are offered
CUHK Agent Value Model	Counts how often each value is offered. The utility of a bid is the sum of the score of its values divided by the best possible score. The model only uses the first 100 unique bids for its estimation
Theoretical Baselines	
Opposite Model	Defines the opponent's utility as one minus the agent's utility
Perfect Model	Perfect knowledge of the opponent's preferences
Worst Model	Defines the estimated utility as one minus the real utility

1. *Bayesian models* estimate the opponent's preferences by first generating a set of candidate preference profiles. Next, Bayesian learning is used to continually update the model, based on certain assumptions about the opponent's concession function. For these models, we also include *Perfect* variants, which use perfect knowledge about the opponent's concessions, but are still unaware of the opponent's exact preferences.
2. *Frequency models* estimate both the issue and value weights separately. The issue weights are estimated based on how often their value is changed between sequential bids. The value weights are derived from the frequency they are offered.
3. *Value models* are similar to the frequency models, except that the issue weights are assumed to be equal.
4. *Theoretical baselines* are used to compare the quality of the models. The *Perfect Model* and *Worst Model* act as an upper and lower bound on quality respectively, while the *Opposite Model* functions as a baseline, since it serves as a good initial guess of the opponent's preferences.

Using the decoupling technique of our BOA architecture (Chap. 3), each model was isolated from existing negotiation agents (as indicated in Table 7.1), and then generalized to be compatible with any bidding strategy. The advantage of using the BOA framework is that we can interchange the opponent modeling component of each negotiation strategy, so that we can compare the performance of different opponent models while keeping the bidding and acceptance strategy fixed.

7.2.3 Selection of Accuracy Measures

We compare the accuracy of opponent models by evaluating how well the models estimate the opponent's preferences when provided with various negotiation traces. In effect, we treat the opponent model as an isolated BOA component that receives offers as input, and yields an estimate of the opponent's preference profile as output, which (hopefully) gets increasingly accurate with every processed bid.

When we assess opponent model accuracy, we require an accuracy measure that quantifies the similarity between the opponent's actual preference profile u_{op} and the estimation u'_{op}. We employed all accuracy measures currently in use (see also Sect. 2.4.4), as shown in Table 7.2. The first two sets of measures are derived from literature, to which we have added a set of metrics based on the Pareto optimal frontier.

We briefly review the definitions of Sect. 2.4.4 of the accuracy measures for outcome spaces and issue weights. Many of the measures are formulated in terms of the combined properties of the opponent's utility space and the agent's own utility space—together called the *bid space*. The *real bid space* B is defined as

$$B = \left\{ \left(u_{own}(\omega), u_{op}(\omega) \right) \mid \omega \in \Omega \right\}.$$

Table 7.2 Overview of accuracy measures

Outcome Space	
Pearson correlation of bids [6]	Pearson correlation coefficient between real and estimated preferences
Ranking distance of bids [4, 6]	Ranking distance between real and estimated preference
Average difference between bids	Average difference between the real and estimated utility of all bids
Issue Weights	
Pearson correlation of issue weights [6]	Pearson correlation coefficient between real and estimated issue weights
Ranking distance of issue weights [6]	Ranking distance between real and estimated issue weights
Average difference between issue weights [7]	Average difference between the real and estimated issue weights
Pareto Frontier	
Average difference of Pareto frontier	The average difference between the real and estimated utility of the Pareto bids
Percentage of found Pareto bids	Percentage of real Pareto bids that are also estimated to be a Pareto bid
Percentage of correct Pareto bids	Percentage of estimated Pareto bids that are also real Pareto bids
Difference in Pareto frontier surface	Absolute difference in surface under the real and estimated Pareto frontier

The *estimated bid space* B' is defined in terms of the estimated opponent utility function u'_{op}:

$$B' = \left\{ \left(u_{own}(\omega), u'_{op}(\omega) \right) \mid \omega \in \Omega \right\}.$$

To quantify how well u'_{op} approximates the opponent's preferences, we might consider the differences between u'_{op} and u_{op} directly. Alternatively, we can analyze the resulting bid spaces B and B', or we might concentrate on subsets.

Outcome space accuracy measures quantify the difference between u_{op} and u'_{op} by considering all bids in the outcome space Ω. A straightforward measure is the *average distance between bids* metric, which calculates the average absolute difference between u_{op} and u'_{op} over Ω. However, as models are usually only concerned with the ranking of outcomes, a more suitable metric is the *Pearson correlation of bids* that measures the correlation between two outcome spaces, which is defined as follows:

$$d_p(u_{op}, u'_{op}) = \frac{\sum_{\omega \in \Omega} (u_{op}(\omega) - \overline{u_{op}})(u'_{op}(\omega) - \overline{u'_{op}})}{\sqrt{\sum_{\omega \in \Omega} (u_{op}(\omega) - \overline{u_{op}})^2 \sum_{w \in \Omega} (u'_{op}(\omega) - \overline{u'_{op}})^2}}, \tag{7.2}$$

where $\overline{u_{op}}$ and $\overline{u'_{op}}$ denote the real and estimated average utility over all outcomes. Alternatively, the *ranking distance of bids* compares all pairwise preference orderings:

$$d_r(u_{op}, u'_{op}) = \frac{1}{|\Omega|^2} \sum_{\omega \in \Omega, \omega' \in \Omega} c_{\prec u, \prec u'}(\omega, \omega'),\qquad(7.3)$$

where $c_{\prec u, \prec u'}$ is the conflict indicator function, which is equal to one when the ranking of the outcomes ω and ω' differs between the two profiles, and zero otherwise.

Issue weight accuracy measures quantify the difference between the issue weights of u_{op} and u'_{op}. The underlying idea is that these variables are most important to estimate correctly. The metrics used are identical to the metrics above.

Pareto frontier accuracy measures focus only on the similarity between two Pareto frontiers. This is more challenging since their sizes can be different, so two sets of Pareto frontiers cannot be compared in the same way as outcome spaces or weight vectors. The *average difference of Pareto frontier* metric calculates the difference in utility over all Pareto bids of bid space B. The *percentage of found Pareto bids* measure gives the percentage of Pareto optimal bids of space B that are also in B'. Conversely, the *percentage of correct Pareto bids* metric yields the percentage of Pareto optimal bids in B' that are correct (i.e., in B).

Finally, we introduce the *difference in Pareto frontier surface* measure, which is defined as follows: we take all outcomes in Ω that form the estimated Pareto frontier in B'; we then map these points onto B. Finally, we compute the absolute difference in surface below these points and the actual Pareto frontier, as shown in Fig. 7.1.

7.2.4 Quantifying the Estimation Accuracy

The goal of our first experiment is to quantify the accuracy of opponent models, both in different domains and through time. We outline in detail below the factors of the experimental setup that we believe are important to consider.

7.2.4.1 Influence of the Opponent Model on the Opponent's Actions

When an agent uses an opponent model, it indirectly influences the opponent in two ways: first, a model may *influence the time of agreement*, as a more accurate model

Fig. 7.1 Visualization of the difference in Pareto frontier surface

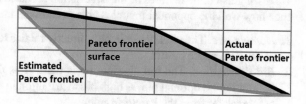

may lead to better offers, resulting in earlier agreement; second, a model may cause changes in the *opponent's strategy*.

Both factors influence the bids presented by the opponent, and thus the information available for the model. To ensure every model learns from the same information, and therefore can be compared with the others, we selected non-adaptive opponents that never accept a bid.

7.2.4.2 Influence of the Opponent's Strategy on the Opponent Model

Opponents differ in how well their behavior corresponds to a model's assumptions. For instance, a model that assumes that the opponent concedes will likely have problems modeling a very competitive agent. Therefore, we should select a balanced set of opponents to avoid favoring any model. One of the defining factors here is how much information an opponent reveals over time. For example, a conceding opponent reveals more of its preferences than an agent who makes random bids. Furthermore, we should include agents that strongly violate the modeling assumptions as to evaluate the robustness of the models. Taking both factors into account, we selected the following agents:

1. *Conceding agents* are time-dependent agents (see Sect. 2.3.3 and Faratin et al. [11]) that select a bid depending on the current time $t \in [0, 1]$ according to a target utility of the form $u_t = P_{max} \cdot (1 - t^{1/e})$ [11]. We selected four agents with $P_{max} = 1$ and different concession rates $e \in \{0.1, 0.2, 1.0, 2.0\}$. These agents make up the *predictable* opponents.
2. *Random agents* offer a random bid above a target utility m, where we selected $m \in \{0, 0.25, 0.50, 0.75\}$. This type of agent and the others below form the *unpredictable* opponents.
3. *Conceding agents with an offset* are time-dependent agents that do not start with their best bid. For this category we use a linear concession rate ($e = 1$) and starting point $P_{max} \in \{0.7, 0.8, 0.9\}$.
4. *Non-conceding agents* start with a minimum target utility that increases to the maximum over time. The target utility is calculated as follows: $u_t = P_{min} + (1 - P_{min}) \cdot t$. We use four agents with parameters $P_{min} \in \{0, 0.25, 0.50, 0.75\}$.

7.2.4.3 Influence of the Scenario on the Opponent Model

We distinguish three features of the negotiation scenario that can significantly influence how well the opponent model is able to estimate the opponent's preferences:

1. *Domain size.* The total possible bids directly relates to the amount of parameters of the preference profile.
2. *Bid distribution.* The bid distribution is defined as the average distance to the nearest Pareto optimal bid. A high bid distribution indicates a high percentage of outcomes far from the Pareto frontier.

3. *Opposition*. The opposition is defined as the distance from the Kalai-Smorodinsky point to the point of perfect satisfaction (maximum utility for both parties). The higher the opposition is, the more competitive the domain.

We refer to Sect. 2.2.4 for the formal definitions. We made sure to select a balanced set of scenarios that display all characteristics. We chose five domains based on their size (see Tables C.2 and D.2): *Itex versus Cypress* (small: 180 bids), *Employment contract* [12] (small: 3125 bids), *Car* (medium: 15625 bids), *Supermarket* (large: 98784 bids), and *Travel* (large: 188160 bids). For each domain we created a set of scenarios varying in bid distribution and opposition. As we defined three levels of degree for both factors, 45 scenarios are used in total.

For the experiment, we ensured that each model processes exactly the same opponent traces, using a maximum amount of $N = 5000$ rounds. For the three groups of deterministic agents we recorded their (unique) negotiation trace, amounting to a total of 495 unique traces. For the random agents we recorded five different traces per agent, thus 900 traces in total. Combined, this amounts to 1395 traces that were used to train every opponent model.

7.2.5 Quantifying the Accuracy/Performance Relationship

The goal of the second experiment is to investigate the relation between accuracy and performance measures, thereby answering our final two research questions. For this experiment, we used a realistic set of opponents whose acceptance strategies are enabled. With realistic opponents, every negotiation is unique, so for this investigation we had to scale down the experimental setup. We selected a set of bidding strategies and scenarios where using a good opponent model would have added value; i.e., tough bidding strategies with limited learning capabilities (i.e., no opponent model), and large, competitive negotiation scenarios. We selected four of the top bidding strategies from ANAC: *Agent K2, HardHeaded, IAMhaggler2011*, and *The Negotiator*; and four time-dependent agents with concession rate $e \in \{0.1, 0.2, 1.0, 2.0\}$. These eight bidding strategies were combined with all thirteen models (the models in Table 7.1 and the two Bayesian models with perfect strategy knowledge) and no model. Each agent competed five times against all opponents (the eight bidding strategies without model) on five scenarios: *Grocery, Employment contract* [12], *Travel, Energy Small*, and *Supermarket*. The first scenarios were used in ANAC 2010 and 2011 (see Tables C.2 and D.2) and the last two scenarios were first used in ANAC 2012; see Sect. E.1.

Each agent played both sides of the five scenarios using a round-based protocol of 1000 rounds. Since 112 agents competed 5 times against 8 opponents on 5 scenarios for both preference profiles, 44800 matches were ran in total.

7.3   Experimental Analysis

We will now answer our three research questions by analyzing the results of both experiments. Each section corresponds to one of the research questions.

7.3.1   Evaluating the Estimation Accuracy of Opponent Models

As outlined in Sect. 7.2.4, we measured the accuracy of a large set of opponent models to answer our first research question, the results of which are shown in Figs. 7.2, 7.3, 7.4, 7.5 and 7.6.

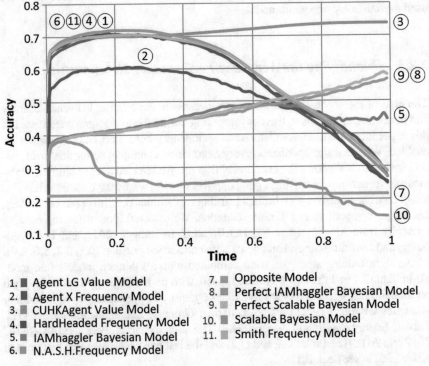

1. ■ Agent LG Value Model
2. ■ Agent X Frequency Model
3. ■ CUHKAgent Value Model
4. ■ HardHeaded Frequency Model
5. ■ IAMhaggler Bayesian Model
6. ■ N.A.S.H.Frequency Model
7. ■ Opposite Model
8. ■ Perfect IAMhaggler Bayesian Model
9. ■ Perfect Scalable Bayesian Model
10. ■ Scalable Bayesian Model
11. ■ Smith Frequency Model

Fig. 7.2 Accuracy over time (measured with *Pearson correlation of bids*) against predictable (i.e., conceding) opponents. The numbers next to a cluster of lines are ordered from high to low accuracy

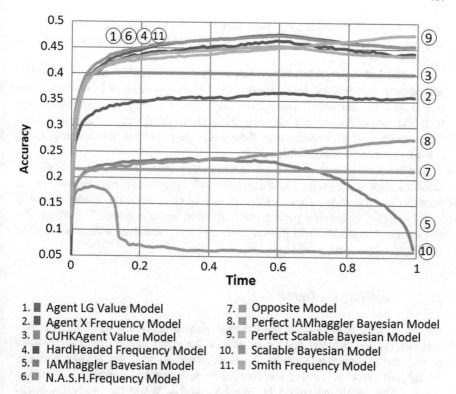

Fig. 7.3 Accuracy over time (measured with *Pearson correlation of bids*) against the unpredictable opponents (i.e., non-conceding agents). The numbers next to a cluster of lines are ordered from high to low accuracy

7.3.1.1 Accuracy over Time

The graphs in Figs. 7.2 and 7.3 show the average accuracy of the opponent models over time. Figure 7.2 shows the accuracy over time when playing against *predictable* opponents; Fig. 7.3 shows the results again the *unpredictable* opponents.

First of all, it is surprising to see that many of the state of the art models actually become *less* accurate over time. The main cause of this phenomenon is that the bids presented later on in the negotiation are incorrectly handled. The value models and frequency models for example, treat every received bid the same way, independent of the time it is received. In effect, this means that when the opponent is conceding, the models increase the estimated utility of less preferred outcomes. This does not hold for the *CUHK Agent Value Model*, which incidentally also performs best, as this model only takes the first 100 unique bids into account. For Bayesian models, the problem is that they assume very particular opponent behavior, which is likely to become increasingly invalid as time progresses, and they perform very poorly as a result. When we disregard this shortcoming by considering the *perfect* Bayesian models, they perform better, and their accuracy then increases monotonically over time. However, even in this case, they come second to the *CUHK Agent Value Model* by a large margin.

Another interesting result is that despite their simplicity, the frequency models and value models perform best against both types of opponents. We believe that this due to the small number of assumptions they make; i.e., only assuming that values with high utility are offered relatively more often. As the right graph illustrates, these models are rather robust, even though it is clear from the final accuracy that it is harder to model unpredictable agents.

The lesson to take away from this is that to be robust, opponent models need to minimize their assumptions about the opponent's behavior. Of course, every model needs to make certain educated guesses, but when it does, the model should at least be highly *adaptable*, paying close attention to the opponent's strategy. The predictions should be revised if, over time, the opponent behavior does not seem to fit the assumptions anymore.

7.3.1.2 Accuracy per Opponent

We now analyze in more detail the accuracy of the best performing models in every category against different opponents; that is, the best value model (*CUHK Value Model*), the best frequency model (*Smith Frequency Model*), and the best performing Bayesian models (*Perfect Scalable Bayesian Model* and *IAMhaggler Bayesian Model*). The results are shown in Figs. 7.5 and 7.4. While the best value model perform best on average, there is no opponent model that dominates all others.

An interesting result is that the technique of the best value model to only take a limited amount of bids into account does not always pay off. The model performs poorly against the non-conceding agents, who show their most preferred values later

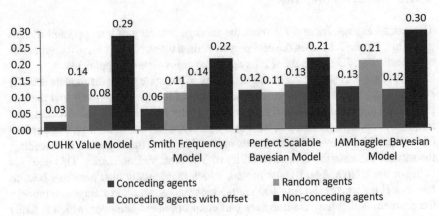

Fig. 7.4 Accuracy of the best four opponent models against different types of opponents, measured using the *Pearson correlation of bids* measure

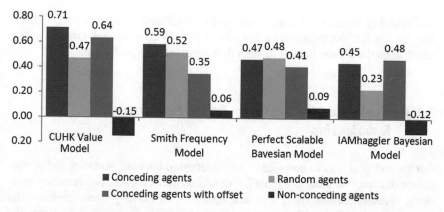

Fig. 7.5 Accuracy of the best four opponent models against different types of opponents, measured using the *average difference of Pareto surface* measure

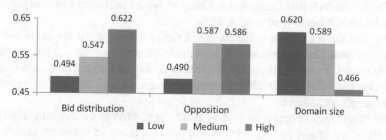

Fig. 7.6 The average accuracy of the best four opponent models on varying scenarios

in the negotiation. This means that the model can be fooled, which can be a concern in practice.

7.3.1.3 Accuracy per Scenario

We are also interested in exactly how the specifics of the negotiation scenario influences accuracy. We focus on the same four opponent models as above, and evaluate their accuracy against predictable opponents. Figure 7.6 summarizes the results. Note that we consider the average accuracy over all four models here, but we have verified that our conclusions also hold for each model individually.

One of the first observations is that there is a high variance in accuracy over different scenarios, and each factor seems to be equally important to consider. This underlines the importance of using a balanced set of negotiation scenarios. Clearly, *domain size* is a significant factor, as the domain size relates directly to the amount of unknown variables to be learnt. But also for the *bid distribution* and *opposition* we find a strong correlation with learning accuracy. The reasons for both are very similar: when the bid distribution or opposition is low, there are many outcomes of similar utility because the average distance to the Pareto frontier is small. This in turn, entails that the values of an issue are relatively close to each other in utility, which is harder for the models to learn than more extreme preferences.

7.3.2 Evaluating the Accuracy Versus Performance Relationship

Our second goal was to investigate the relationship between accuracy and performance of opponent models. Figure 7.7 visualizes the results for two accuracy measures: *Pearson correlation of bids* and *difference in Pareto frontier surface*. The performance is expressed in terms of obtained utility by the agents that employ the opponent models, normalized such that the *Worst model*'s performance is zero, and the *Perfect model*'s performance is 1. Using *no model* falls somewhere in between, since this is still better than using a wrong model.

The first thing to notice is the cluster of the best performers: the value and frequency models. The performance of these models is already quite close to that of the perfect model. To put it differently, we cannot anticipate a significant improvement from any other preference modeling technique over what is already achieved by these rather simple techniques.

The other types of opponent models also form clusters in the diagram. The *(Perfect) Bayesian Model* perform even worse than not using an opponent model; and only slightly better than simply assuming opposite preferences.

The almost linear relationship between accuracy and performance is the second thing that stands out. This shows that there is always added value to increasing the accuracy of an opponent model, even when the accuracy is already high. Nevertheless,

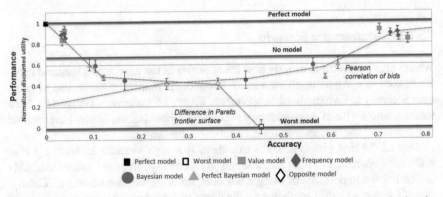

Fig. 7.7 Accuracy versus performance for all opponent models. Accuracy is measured using the *difference in Pareto frontier surface* (range [0, 1], where 0 is best) and *Pearson correlation of bids* (range [−1, 1], where 1 is best)

the added value will necessarily be small, as the performance is already at 90 % of its upper limit when the accuracy is at 70 %.

An interesting comparison can be made with the results of the previous experiment. Figure 7.2 clearly shows the decrease in accuracy over time of many of the frequency and value models. How is it that they still manage to perform close to optimally? The reason is that many negotiations end in agreement, and this occurs somewhere *before* the deadline by definition. In these cases, the models are updated with less bids of poor value for the opponent. Therefore, the deciding factor in the success of the value and frequency models lies in their higher initial accuracy.

Finally, it is interesting that the results for the *Pearson correlation of bids* and *difference in Pareto frontier surface* metrics are in fact very similar when we ignore their orientation. Despite that the latter only measures the quality of the Pareto optimal frontier instead of the full outcome space, it seems to be a suitable predictor for performance as well. We explore this idea further in the next paragraph.

7.3.3 Evaluating the Usefulness of Accuracy Measures

Our final goal was to find a strong predictor for performance of opponent models, since there are so many different accuracy measures to pick from. Towards this end, we applied all of the accuracy measures shown in Table 7.2 and analyzed their correlation with performance; see Fig. 7.8.

The dark line represents the predictive power of each accuracy measure, which is defined by the absolute correlation coefficient $|\rho|$ between the accuracy measure score and the model's performance. We take the absolute value because some accuracy measures are negatively correlated with performance, while others are positively correlated.

The light gray line indicates what portion of the bid space is learned by each accuracy measure. For this, we calculate the absolute correlation coefficient $|\rho|$ between each accuracy measure and the *Pearson correlation of bids*. At the lower end of the scale we see the accuracy measures that only consider issue weights, which means they are not correlated at all with learning the space as a whole. These measures should not be used to make predictions about performance because they do not convey enough information about the accuracy of a model.

We found three measures that correlate strongly with performance, as indicated by the dark gray line, and therefore are good performance predictors; these are: *difference in Pareto frontier surface*, *Pearson correlation of bids*, and *Ranking distance of bids*. These measures codify sufficient information about the relationship between the real preferences and the learned preferences, and therefore, we can translate these notions to statements about performance. The performance of the top three measures are significantly better than the other measures (*one-tailed t-test, $p < 0.01$*).

Even though it only quantifies the similarity of the Pareto frontier, the *difference in Pareto frontier surface* metric performs best of all (*one-tailed t-test, $p < 0.02$*). This means that for an opponent model, it is sufficient to predict which bids are

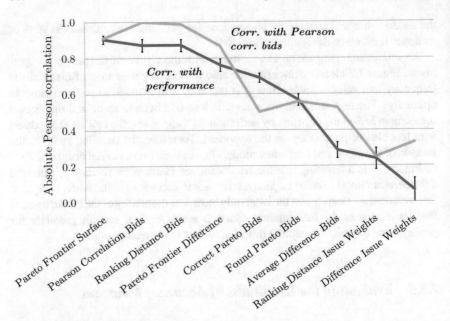

Fig. 7.8 Absolute correlation between accuracy measure scores and two other measures: performance and *Pearson correlation of bids*

Pareto optimal. The reason being that the Pareto frontier is a crucial component of the outcome space, and that many bidding strategies seek Pareto optimal agreements. It is less computationally expensive to calculate than the *ranking distance of bids*, and it has the lowest standard deviation between the runs. Furthermore, it is defined for all inputs, in contrast to the Pearson correlation measures, whose results are undefined when all bids are estimated to have the same utility. Therefore, we recommend the *difference in Pareto frontier surface* as a suitable measure for accuracy.

7.4   Conclusion

In this chapter we evaluated a large set of accuracy measures to identify the best method to predict the performance of opponent modeling techniques in negotiation. We introduced a procedure to quantify the accuracy of state of the art opponent models and we identified their strengths and weaknesses. One of our main conclusions is that there is an almost linear correspondence between accuracy and performance of models when we employ the proper accuracy measures. Moreover, the best models are close to being perfectly accurate, which means there is only limited room for improvement with regard to performance.

Surprisingly, the accuracy of most opponent models decays over time due to the incorrect handling of the opponent's less preferred bids, which are usually offered at a later stage of the negotiation. Especially then, good strategy prediction is needed in order to be effective at preference modeling.

Finally, we analyzed how well accuracy measures can predict the performance of an opponent model. Three measures in particular are useful predictors of performance, and this can be best achieved by limiting the analysis to difference in Pareto frontier surface between the real and the learned bid space.

This chapter is based on the following publications: [13]

Tim Baarslag, Mark J.C. Hendrikx, Koen V. Hindriks, and Catholijn M. Jonker. Predicting the performance of opponent models in automated negotiation. In *International Joint Conferences on Web Intelligence (WI) and Intelligent Agent Technologies (IAT), 2013 IEEE/WIC/ACM*, volume 2, pages 59–66, Nov 2013

References

1. Coehoorn RM, Jennings NR (2004) Learning an opponent's preferences to make effective multi-issue negotiation trade-offs. In: Proceedings of the 6th international conference on electroniccommerce, ICEC '04, ACM, New York, USA, pp 59–68
2. Zeng D, Sycara KP (1998) Bayesian learning in negotiation. Int J Hum-Comput Stud 48(1):125–141
3. Hindriks KV, Tykhonov D (2008) Opponent modelling in automated multi-issue negotiation using bayesian learning. In: Proceedings of the 7th international joint conference on autonomous agents and multiagent systems, international foundation for autonomous agents and multiagent systemsvol 1 of AAMAS '08, Richland, SC, pp 331–338
4. Buffett S, Spencer B (2005) Learning opponents' preferences in multi-object automated negotiation. In: Proceedings of the 7th international conference on electronic commerce, ICEC '05, ACM, New York, USA, pp 300–305
5. Carbonneau RA , Kersten GE, Vahidov RM (2008) Predicting opponent's moves in electronic negotiations using neural networks. Expert Syst Appl, 34(2):1266–1273
6. Hindriks KV, Tykhonov D (2010) Towards a quality assessment method for learning preference profiles in negotiation. In: Ketter W, La Poutré JA, Sadeh N, Shehory O, Walsh W (eds) Agent-mediated electronic commerce and trading agent design and analysis, Lecture notes in business information processing, vol 44. Springer, Berlin, pp 46–59
7. Jazayeriy H, Azmi-Murad M, Sulaiman N, Udizir NI (2011) The learning of an opponent's approximate preferences in bilateral automated negotiation. J Theor Appl Electron Commer Res, 6(3):65–84
8. Williams CR, Robu V, Gerding EH, Jennings NR (2012) Iamhaggler: a negotiation agent for complex environments. In: Ito T, Zhang M, Robu V, Fatima S, Matsuo T (eds) New trends in agent-based complex automated negotiations. Stud Comput Intell, Springer, Berlin, pp 151–158
9. van Krimpen T, Looije D, Hajizadeh S (2013) Hardheaded. In: Ito T, Zhang M, Robu V, Matsuo T (eds) Complex automated negotiations: theories, models, and software competitions, Stud Comput Intel, vol 435. Springer, Berlin, pp 223–227
10. van Galen Last N (2012) Agent Smith: Opponent model estimation in bilateral multi-issue negotiation. In: Ito T, Zhang M, Robu V, Fatima S, Matsuo T (eds), New trends in agent-based complex automated negotiations. Stud Comput Intel, Springer, Berlin, pp 167–174
11. Faratin P, Sierra C, Jennings NR (1998) Negotiation decision functions for autonomous agents. Robot Auton Syst 24(3–4):159–182
12. Thompson L (2000) The Mind and heart of the negotiator, 3rd edn. Prentice Hall Press, Upper Saddle River, NJ, USA

13. Baarslag T, Hendrikx MJC, Hindriks KV, Jonker CM (2013) Predicting the performance of opponent models in automated negotiation. In: International joint conferences on web intelligence (WI) and intelligent agent technologies (IAT), 2013 IEEE/WIC/ACM, vol 2, pp 59–66, Nov 2013

Chapter 8
A Quantitative Concession-Based Classification Method of Bidding Strategies

Abstract In this chapter, we cover the last agent strategy component of the BOA architecture of Chap. 3, namely the *bidding strategy*; i.e., the strategy component that decides the *concessions* to be made during the negotiation. Every negotiator needs to make concessions to successfully reach an agreement, and the willingness to do so depends in large part on the opponent. A concession by the opponent may be reciprocated, but the negotiation process may also be frustrated if the opponent does not concede at all. This process of concession making is a central theme in many automated negotiation strategies. In this chapter, we present a quantitative *classification method* of negotiation strategies that measures the willingness of an agent to concede against different types of opponents. We classify some well-known negotiating strategies with respect to their concession behavior, including the ANAC agents we described in Appendix B. We show that our technique makes it easy to identify the main characteristics of negotiation agents, and that it can be used to group negotiation strategies into four categories with common negotiation characteristics, namely *Inverter*, *Conceder*, *Competitor*, and *Matcher*. We are able to conclude, among other things, that different kinds of opponents call for adopting a different negotiation orientation. Our analysis allows us to highlight several interesting insights for the broader automated negotiation community. In particular, we show that the most adaptive negotiation strategies are not necessarily the ones that win the competition.

8.1 Introduction

When two agents are conducting a negotiation, the opening offer and first counteroffer define the initial bargaining range [7] of the negotiation. Sometimes the other party will immediately accept the offer, or will state that the set of demands is unacceptable, breaking off the negotiation. But usually, after the first round of offers, the question is: what concession is to be made next? One can choose to signal a position of firmness and stick to the original offer. Or one can take a more cooperative stance, and choose to make a concession. If one side is not prepared to make concessions, the other side must capitulate, or more commonly, the negotiation will end up in a break off.

© Springer International Publishing Switzerland 2016
T. Baarslag, *Exploring the Strategy Space of Negotiating Agents*,
Springer Theses, DOI 10.1007/978-3-319-28243-5_8

Making concessions is key to a successful negotiation: without them, negotiations would not exist [7].

Negotiation can even be defined in terms of making concessions: Pruitt [8] defines it as a process by which a joint decision is made by two or more parties that first verbalize contradictory demands and then move towards agreement by a process of concession making or search for new alternatives.

Many of the classic negotiation strategies are characterized by the way they make concessions throughout the negotiation. For example, the time-dependent tactics such as *Boulware* and *Conceder* [4] are characterized by the fact that they steadily concede throughout the negotiation process. Concessions made by such tactics depend on various factors, such as the opening bid, the reservation value, and the remaining time. Where possible, it can be beneficial to use prior information about the negotiation setting in selecting a suitable concession strategy. However, when there is no such prior information available to the agents, deciding what concessions to make depends in large part on the opponent. One can either choose to signal a position of firmness and stick to an offer. Alternatively, one can take a more cooperative stance, and choose to make a concession. Such a concession may, in turn, be reciprocated by the opponent, leading to a progression of concessions. On the other hand, the negotiation process can easily be frustrated if the opponent adopts a take-it-or-leave-it approach.

Against this background, we study the ANAC negotiation strategies according to the way they concede towards different types of opponents. The competition results of ANAC (Tables C.2 and D.2 of the Appendix) only give a fairly narrow view of the performance of the agents. In order to consider their behavior in more detail, and also to explore the interactions between different strategies in a tournament setting, we use a more sophisticated evaluation method in this chapter. As will become evident, the performance of an agent depends heavily on the opponent.

The work in this chapter advances the state-of-the-art in automated negotiation in the following ways. We present a new classification method for negotiation strategies, based on their pattern of concession making when faced with different kinds of opponents. We introduce a definition of Concession Rate (CR) which measures the cooperativeness of an agent. We present a technique to quantitatively measure the CR against two extreme types of strategies: a take-it-or-leave-it strategy, and a conceding tactic. We then apply this technique to classify some well-known negotiating strategies, including the agents of ANAC. We present an in-depth analysis of the strategies from the finalists of ANAC and the techniques employed by the different agents. This provides, for the first time, insight into the negotiation strategy space of the ANAC agents and what type of behavior (e.g. being adaptive or hardheaded) is more likely to do well in different situations. It also aids our understanding of what concession making strategies are effective in settings such as ANAC.

Analyzing the competition results yields some interesting insights into the properties exhibited by agents which successfully negotiate in realistic negotiation environments. We compare, for each strategy, the correlation between the total amount of concession made during the negotiation, and the utility it achieves. Among other things, we observe that different kinds of opponents call for a different approach in making concessions. For instance, a successful negotiating agent should behave competitively, especially against very cooperative strategies. We find that there is a direct correlation between the performance of a strategy and its concession rate against a simple *Conceder Linear* strategy, but there is less correlation between performance and concession against a non-concessive *Hardliner* strategy. This shows that the best strategies try to exploit concessive opponents, but against a non-concessive opponent, some aim to reach any agreement (even with a low utility) while others prefer not to concede and may therefore may receive the disagreement payoff (which is zero). Moreover, we conclude that our technique has the desirable property of grouping negotiation strategies into categories with common negotiation characteristics.

The remainder of this chapter is organized as follows. Section 8.2 provides an overview of concession making in negotiation, including our adopted model of negotiation. In Sect. 8.3 we give the definition of concession rate, and we outline a method it. This is followed by Sect. 8.4 that presents our experimental results, and collects the insights gathered by this analysis. Finally, Sect. 8.5 presents this chapter's conclusions.

8.2 Concession Making in Negotiation

In earlier work on conflict management through negotiation, the negotiation stance is characterized by two orientations: cooperative and competitive [3]. The theory relates to two basic types of goal interdependence that negotiators might have. It is either positive, where the negotiators' goals are linked in such a way that their goal attainments are positively correlated ('sink or swim together'), or the interdependence is negative, namely when the goal attainments are negatively correlated ('when one swims, one sinks').

However, a negotiator's stance is usually not limited to one of the two orientations, because negotiation is a dynamic process and the position of the negotiators can change in response to the other party's information or behavior [7]. In this chapter, we take the stance that negotiators can exhibit a mixture of the two orientations, mainly depending on the type of opponent (see Fig. 8.1). For example, a negotiator may cooperate with a cooperative opponent, but the same negotiator may be very competitive when facing competition. That is, in this case it *matches* the behavior of the opponent.

Conversely, a negotiator can be cooperative towards a competitive opponent and at the same time exploit cooperative opponents by playing competitive against them. In that case, it *inverts* the opponent's behavior.

This way, we distinguish four types of negotiation orientations depending on the behavior against the opponent (see Table 8.1): *Inverter*, *Conceder*, *Competitor*, and *Matcher*. Every negotiation orientation corresponds to a different stance towards either of the two types of opponents. One of the main contributions of this chapter is to define a formal, mathematical procedure for classifying agents into one of the four categories.

Fig. 8.1 The diagram of conceding behavior against cooperative and competitive opponents

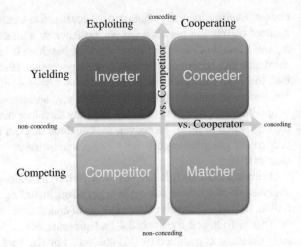

Table 8.1 Four types of negotiation orientations

Orientation	versus Conceder	versus Hardliner
Inverter	Exploiting	Yielding
Conceder	Cooperating	Yielding
Competitor	Exploiting	Competing
Matcher	Cooperating	Competing

8.3   Concession Rate

In this section we introduce the notion of concession rate, which quantifies the amount an agent has conceded towards the opponent during a negotiation. We will define the concession rate of an agent A as a normalized measure $CR_A^t \in [0, 1]$ with the following meaning: if $CR_A^t = 0$, then A did not concede at all up to time point t, while $CR_A^t = 1$ means that player A yielded completely to the opponent by offering the opponent its best offer.

Note that it is generally not enough to simply consider the utility of the agreement as a measure for the concession rate. For instance, a negotiator may not get an agreement before the deadline. In that case, both parties receive zero utility, but this gives no information about the concessions that were made. Also, the last offer made by a negotiator is not necessarily the offer to which he was ultimately willing to concede. To capture the notion of concession rate, we define it in terms of the *minimum* utility m a negotiator has demanded during the negotiation, as this is a measure of the total amount the negotiator was willing to concede.

We define the concession rate CR_A^t for arbitrary time t, but our work only deals with the concession rate of a player A at the *end* of the negotiation thread. We shall denote this simply by CR_A. We also omit the subscript A when it is clear from the context.

Fig. 8.2 The three possible outcomes of the *Nice or Die* scenario, with the utility of agent A plotted on the *horizontal* axis

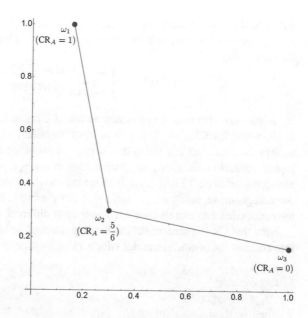

8.3.1 An Example

We illustrate the concept of CR by considering the *Nice or Die* scenario as described in Section D.2 of the Appendix. *Nice or Die* has only three possible outcomes: $\omega_1 = \langle 0.16; 1 \rangle$, $\omega_2 = \langle 0.3; 0.3 \rangle$, and $\omega_3 = \langle 1; 0.16 \rangle$ (see Fig. 8.2).

Let us first consider the case where agent A sticks to the same offer ω_3 throughout the negotiation, demanding the highest possible utility for itself. In this case, the minimum utility m that player A has demanded is equal to 1, so A has not conceded anything, therefore the corresponding concession rate of agent A should be equal to zero:

$$CR_A = 0, \text{ for } m = 1. \tag{8.1}$$

On the other hand, when A concedes all the way to the opponent's best option (which is ω_1), CR_A should be equal to 1. Note however, that conceding all the way does not necessarily mean demanding zero utility, or the lowest possible utility. For instance in this example, A would still receive 0.16 utility when bidding ω_1 (see Fig. 8.2). In general, there may also be bids with even lower utility for A that are, for example, also bad for its opponent. However, player A should be able to always obtain at least an agreement that is the best outcome for the opponent, as any rational player B will accept it. We shall refer to this utility as the *full yield utility* (FYU$_A$) of player A. Intuitively, it is the worst score player A can expect to obtain in the negotiation. Player A's concession rate is therefore maximal if he makes a bid on or below the full yield utility:

$$CR_A = 1, \text{ for } m \leq FYU_A. \tag{8.2}$$

Between the two extremes, we define CR_A to decrease linearly from 1 to 0. There is only one function that satisfies this constraint, along with the Eqs. (8.1) and (8.2) above, namely:

$$CR_A(m) = \begin{cases} 1 & \text{if } m \leq FYU_A, \\ \frac{1-m}{1-FYU_A} & \text{otherwise.} \end{cases}$$

By using normalization, it is guaranteed that if $CR_A = 0$, then A has not conceded at all, while for $CR_A = 1$, player A has conceded fully (i.e., up to its full yield utility). Normalizing has the added benefit of reducing bias in the scenarios: in a typical scenario with strong opposition such as *Energy*, players may obtain utilities anywhere between 0.1 and 1, while in scenarios with weak opposition such as *Company Acquisition*, utility ranges are much more narrow. Normalization ensures that the concession rate can be compared over such different scenarios.

Note that CR is a measure of the bidding strategy, so in particular it does not take into account the conditions under which an agent accepts an offer.

Example
Suppose player A has made the following bids: $\langle \omega_3, \omega_3, \omega_2, \omega_3 \rangle$ (see Fig. 8.2). Then its minimum demanded utility m is equal to the utility of ω_2, which is 0.3. The full yield utility FYU_A is equal to the utility of ω_1, which is 0.16. Therefore,

$$CR_A = \frac{1 - 0.3}{1 - 0.16} = \frac{5}{6}.$$

8.3.2   Formal Definition

Suppose a player A has a utility function u_A, mapping any outcome in Ω into the range $[0, 1]$. As we have assumed that the utility function is normalized in our setting, there will exist an *optimal outcome* $\overline{\omega}_A \in \Omega^1$ for which $u_A(\overline{\omega}_A) = 1$ (see Sect. 2.2.4). In typical negotiation domains, the corresponding utility $u_B(\overline{\omega}_A)$ of this outcome is far from optimal for player B, because the best outcome for A is typically not the best outcome for B. Player B should be able to always obtain at least this outcome in a negotiation, as A will always be inclined to accept it. We shall refer to this utility as the *full yield utility* (FYU$_B$) of player B (see Fig. 8.3). Intuitively, it is equal to his bottom line utility.

As defined in Sect. 2.2.2, we use the alternating-offers protocol supplemented with a real time line $\mathcal{T}$ with a deadline. We represent by $x^t_{A \to B}$ the negotiation outcome

---

[1]Note that an optimal outcome $\overline{\omega}_A$ is not necessarily unique, but typical domains (including those considered in ANAC and hence, in this chapter) all have unique optimal outcomes for both players, so that the full yield utility is well-defined.

Fig. 8.3 The yield of player A is determined by MIN_A^t

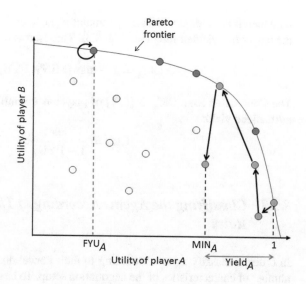

proposed by agent A to agent B at time t. Recall from Sect. 2.2.2 that a *negotiation thread* (cf. [4, 9]) between two agents A and B at time $t \in \mathcal{T}$ is defined as a finite sequence

$$H_{A \leftrightarrow B}^t := \left(x_{p_1 \to p_2}^{t_1}, x_{p_2 \to p_3}^{t_2}, x_{p_3 \to p_4}^{t_3}, \ldots, x_{p_n \to p_{n+1}}^{t_n} \right),$$

where

1. The offers are ordered over time: $t_k \leq t_l$ for $k \leq l$.
2. The offers are alternating between the agents: $p_k = p_{k+2} \in \{A, B\}$ for all k.
3. All t_i represent instances of time $\mathcal{T}$, with $t_n \leq t$,
4. The agents exchange complete offers: $x_{p_k \to p_{k+1}}^{t_k} \in \Omega$ for $k \in \{1, \ldots, n\}$.

For any $t \in \mathcal{T}$, let

$$H_{A \to B}^t = \left\{ x_{A \to B}^s \in H_{A \leftrightarrow B}^t \mid s \leq t \right\}$$

denote all bids offered by A to B until time t in an active negotiation thread. We can now formulate the minimum utility that agent A demanded during the negotiation thread $H_{A \to B}^t$. That is to say, we consider the largest concession the player has made so far:

$$MIN_A^t = \min\{ u_A(x) \mid x \in H_{A \to B}^t \}$$

Informally, MIN_A^t denotes the lowest that A is willing to bid up until time t. The inverse of this is called the *yield* of player A. The lower player B is willing to go, the larger the yield. A yield of zero means the player has made no concession whatsoever (and therefore his demanded utility remains equal to one); A yield of $1 - \text{FYU}$ means the player has yielded fully (see Fig. 8.3). That is, it is defined as:

$$\text{Yield}_A^t = 1 - \max\left(\text{MIN}_A^t, \text{FYU}_A\right).$$

The Concession Rate $\text{CR}_A^t \in [0, 1]$ of player A up until time t is then simply the normalized yield:

$$\text{CR}_A^t = \frac{\text{Yield}_A^t}{1 - \text{FYU}_A}.$$

8.3.3 Classifying the Agents According to Their Concession Rates

In order to classify agents according to their concession rate, we must consider a number of characteristics of the negotiation setup. To be able to compare the results, we need to fix the set of opponents that are used to measure the CR. This raises the question which agents can be used for this purpose. To test both sides of the spectrum, we let the finalists negotiate against both a very cooperative and a very competitive opponent. The opponent tactics that we use to measure concession rates are simple, non-adaptive negotiation tactics. We do so because we want to ensure that the concession rate results depend as much as possible on the agent's own negotiating tactic, and not on the opponent's. To be more precise, we aim for three opponent characteristics when measuring the concession rate:

1. **Simplicity**: The concession rate results of an agent are less sensitive to the opponent, and hence easier to interpret, if the opponent negotiation tactic is simple and easy to understand.
2. **Regularity**: We want to give the agent sufficient time to show its bidding behavior; therefore, the opponent should not end the negotiation prematurely by either reaching an agreement too fast or breaking off the negotiation. Another issue here is that the opponent should generate sufficient bids. This requires computationally efficient agents that respond within a reasonable amount of time and excludes extreme agents that only make a limited number of offers.
3. **Deterministic behavior**: In order to reduce variance in experimental results, we prefer deterministic agents to those that demonstrate random bidding behavior.

For the *competitive opponent*, we chose *Hardliner* (also known as take-it-or-leave-it, or *Hardball* [7]). This strategy simply makes a bid of maximum utility for itself and never concedes. This is the most simple competitive strategy that can be implemented and it fits the other two criteria as well: it is deterministic and it gives the agent the full negotiation time to make concessions.

For the *cooperative opponent*, we selected *Conceder Linear*; i.e., the time-dependent tactic adapted from [4, 5] with parameter $e = 1$ (see also Sect. 2.3.3).

Both strategies accept if and only if their planned offer has already been proposed by the opponent in the previous round.

There exist even simpler conceding tactics such as *Random Walker*[2] (which generates random bids), or an agent that accepts immediately. However, both opponent strategies are not *regular* in the sense that they do not give the agent enough time to show its bidding behavior. *Random Walker* has the additional disadvantage of not being deterministic. Therefore, we believe *Random Walker* can serve as a useful baseline strategy to test the efficacy of a negotiation strategy, but not as a useful opponent strategy to measure an agent's willingness to concede. Consequently, we selected *Conceder Linear* as the cooperative opponent, as it fulfills the three requirements listed above.

8.4 Experiments

When looking at the results of the ANAC 2010 and 2011 competitions listed in Tables C.2 and D.2, it is natural to ask which agent characteristics were decisive factors in the final ranking of the agents. Which agents behaved very competitively and which ones were more cooperative? Were they successful because of it, and if so, against whom? In order to answer such questions, we characterize the ANAC agents by analyzing their bidding behavior and utility gain against different types of opponents. In particular, we focus on the amount they are willing to concede to the opponent, as this is a key determinant of an agent's bidding behavior.

8.4.1 Experimental Setup

For our experimental setup we employed GENIUS to simulate tournaments between negotiating agents, using the same negotiation parameters as described in ANAC (see Appendix B). In our experimental setup we included all the negotiation tactics that were submitted to ANAC 2010 and 2011, which we analyze separately.

In addition to the ANAC agents, we included some well-known agents to explore some extreme cases. First, we included the *Hardliner* strategy described in Sect. 2.3.3, which consistently makes the maximum bid for itself. We also studied three members of the time-dependent-tactics family [4] as defined in Sect. 2.3.3, namely: *Boulware* ($e = 0.2$), *Conceder Linear* ($e = 1$), and *Conceder* ($e = 2$). Finally, we included the *Random Walker* strategy, which randomly jumps through the negotiation space. We shall refer to these five strategies as our *benchmark strategies*. For the ANAC 2010 experiment, we also included a variant of the *Relative Tit for Tat* agent from [4].

---

[2]The *Random Walker* strategy is also known as the *Zero Intelligence* strategy [6].

This strategy, called *Simple Nice Tit for Tat*, tries to reproduce the behavior that its opponent performed in the previous step.

We measured the concession rate of an agent *A* playing against *Conceder Linear* and *Hardliner*. The two parties attain a certain outcome, or reach the deadline. In both cases, at the end of the negotiation, *A* has reached a certain concession rate as defined in Sect. 8.3.2. The concession rate is then averaged over all trials on all scenarios, alternating between the two preference profiles defined as part of that scenario. For example, in the *Energy* scenario, *A* will play both as the electricity distribution company and as the consumer. We then repeated every negotiation 30 times to increase the statistical significance of our results. We have taken the averages and standard deviations over these 30 data points, where each data point is the average taken over all scenarios.

For the ANAC 2010 agents, we picked two out of the three domains that were used in ANAC 2010 (see Table C.2). We omitted the third domain (*Travel*) as some of the ANAC agents did not scale well and had too many difficulties with it to make it a reliable testing domain. For the ANAC 2011 agents, we used all eight scenarios of the finals (see Table D.2).

8.4.2 Experimental Results for ANAC 2010

We present the results of the experiments for ANAC 2010 in Table 8.2 and its graphical representation is depicted in Fig. 8.4.

Table 8.2 An overview of the concession rate of every agent in the ANAC 2010 experiments

Agent	CR versus Conceder	CR versus Hardliner
Agent K	0.12	0.18
Agent Smith	0.46	1.00
Boulware	0.14	1.00
Conceder linear	0.43	1.00
Conceder	0.63	1.00
FSEGA	0.33	0.76
Hardliner	0.00	0.00
IAMcrazyHaggler	0.05	0.05
IAMhaggler	0.02	0.27
Nozomi	0.20	0.22
Random walker	0.97	1.00
Simple nice tit for tat agent	0.42	0.01
Yushu	0.11	0.95

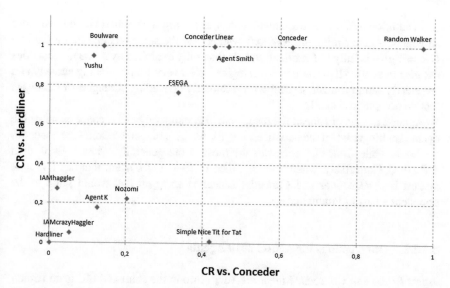

Fig. 8.4 A graphical overview of the concession rates of the ANAC 2010 agents

8.4.2.1 General Observation

We start with the general observations (which also hold for the ANAC 2011 results) regarding the clustering of different strategies in Fig. 8.4. In particular, the *Hardliner* strategy and the *Random Walker* strategy are at the opposite sides of the concession rate spectrum. *Hardliner* will not concede to any type of strategy, so by definition it has $CR = 0$ against both *Hardliner* and *Conceder Linear*. Consequently, *Hardliner* is the least conceding strategy possible.

On the other hand, *Random Walker* will make arbitrary concessions given enough time. This makes *Random Walker* one of the most concessive strategies possible. Against *Hardliner*, it is just a matter of time for *Random Walker* to randomly produce a bid with which it fully concedes, so *Random Walker* has $CR = 1$ against *Hardliner* when it manages to reach an agreement. In large domains, however, it may not be able to produce such a bid in time, as we will see when we discuss the results of ANAC 2011, which featured larger domains. Against *Conceder*, it may not have as much time to fully concede, but it generally will produce offers of very low utility in this case as well, resulting in a CR of 0.97.

We included three members of the time-dependent tactics family in our experiments: *Boulware* ($e = 0.2$), *Conceder Linear* ($e = 1$), and *Conceder* ($e = 2$). They are all located in the top of the chart because they all share the same concession rate of 1 when playing against *Hardliner*. This is to be expected, as any agent from the time-dependent family will offer the reservation value when the deadline is reached [5], resulting in full concession to the opponent. In general, all time-dependent tactics will lie on the line $CR = 1$ against *Hardliner*.

In addition to the time-dependent tactics, at the top of the chart we see two more strategies: *Agent Smith* and *Random Walker*. All the strategies that lie on the line $CR = 1$ give in fully to *Hardliner* and are thus fully exploited by a strategy that does not give in at all. All of these four strategies have a very simple bidding strategy and are clearly not optimized to deal with minimally conceding strategies or strategies that do not concede at all.

Against *Conceder Linear*, the results are also intuitively clear: concessions of the agents get bigger when the parameter $e \in (0, \infty)$ gets bigger, so *Boulware* concedes the least, while *Conceder* concedes the most. More generally, when $e \to 0$, then $CR \to 0$. Conversely, when $e \to \infty$, then $CR \to 1$. An agent that has $CR = 1$ against both *Hardliner* and *Conceder Linear* is an agent that would jump to the opponent's best bid immediately.

8.4.2.2   Observations for ANAC 2010 Agents

Agent Smith and *Conceder Linear* are very close in the chart and this is no coincidence: *Agent Smith* uses essentially the same strategy as *Conceder Linear*, by first making a proposal of maximum utility and subsequently conceding linearly towards the opponent.

The same holds for *Yushu* and *Boulware*: the strategies are very similar, as is indicated by their close vicinity in the chart. Like *Boulware*, *Yushu* adopts a very competitive time dependent strategy, making larger concessions when the negotiation deadline approaches. Both adopt a conservative concession making strategy and are not willing to make large concession at the beginning, but prefer to wait for their opponent to make concessions. These two examples show that this chart can be useful to cluster strategy types, as similar strategies have similar concession characteristics.

Finally, there is a big cluster of strategies in the left part of the chart, which is populated by the top four strategies in ANAC 2010: *Agent K*, *IAMhaggler*, *Nozomi* and *Yushu*. The better performing strategies of ANAC 2010 have different approaches towards the *Hardliner*, but they seem to have one trait in common: they all concede very little to the *Conceder*. In other words: they exploit the *Conceder* by waiting for even bigger concessions. The fact that these strategies did very well in ANAC 2010 seems to indicate that in order to be successful in an automated negotiation competition, an agent should behave competitively, especially against very cooperative strategies. We explore this theme more deeply in our ANAC 2011 experiment.

8.4.2.3   Four Negotiation Orientations

This section makes observations regarding the clustering of different strategies in Fig. 8.4, and then classifies them into the four negotiation orientations of Fig. 8.1. This procedure is necessarily arbitrary; nevertheless, we propose the following grouping.

The top left agents in the diagram can be considered to be *Inverters*: *Yushu*, *Boulware*, and *FSEGA*. The remaining agents in the top right are then *Conceders*, namely: *Conceder Linear*, *Agent Smith*, *Conceder*, and *Random Walker*.

The *Simple Nice Tit for Tat* strategy is the only strategy that can be considered a *Matcher*, i.e.: it does not concede to a *Hardliner*, but it does concede to the *Conceder*. Clearly, this is to be expected from a *Tit for Tat* strategy, as it is based on cooperation through reciprocity: it matches whatever the other player did on the preceding move. The fact that this type of strategy does not occur naturally in ANAC 2010 can be explained by our previous comments on clustering: following a *Tit for Tat* strategy is not as successful in negotiation, because it does not exploit the conceding strategies.

All of the remaining strategies are *Competitors*, i.e. they do not concede much, whether it is against a cooperative or a competitive agent. The majority of strategies that performed well during ANAC are located in this region. Again, we observe that the successful strategies are very competitive.

8.4.3 Experimental Results for ANAC 2011

We present the results of the concession rate experiments for ANAC 2011 in Table 8.3, and its graphical representation is depicted in Fig. 8.5. In Figs. 8.6, 8.7, 8.8 and 8.9, we present the connection between the concession rate of the agents and their average negotiation results against *Conceder Linear* and *Hardliner*. In all figures, the error bars represent one standard deviation from the mean.

Table 8.3 An overview of the concession rate and standard deviation of every ANAC 2011 agent in the experiments against *Conceder Linear* and *Hardliner*

Agent	CR versus Conceder linear		CR versus Hardliner	
	Mean	Standard deviation	Mean	Standard deviation
Agent K2	0.232	0.029	0.435	0.019
BRAMAgent	0.141	0.042	0.642	0.002
Gahboninho	0.219	0.041	0.907	0.032
HardHeaded	0.041	0.027	0.622	0.039
IAMhaggler2011	0.419	0.058	0.987	0.005
Nice tit for tat agent	0.169	0.033	0.876	0.031
The negotiator	0.071	0.011	1.000	0.000
ValueModelAgent	0.239	0.013	0.667	0.030
Random walker	0.848	0.040	0.904	0.030
Hardliner	0.000	0.002	0.000	0.002
Boulware	0.162	0.009	1.000	0.000
Conceder linear	0.414	0.034	1.000	0.000
Conceder	0.571	0.043	1.000	0.000

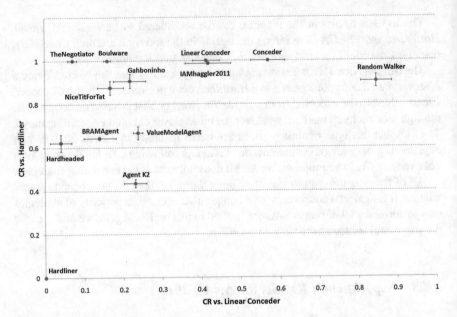

Fig. 8.5 A scatter plot of the concession rate and standard deviation of every ANAC 2011 agent in the experiments against *Conceder Linear* and *Hardliner*. The error bars represent one standard deviation from the mean

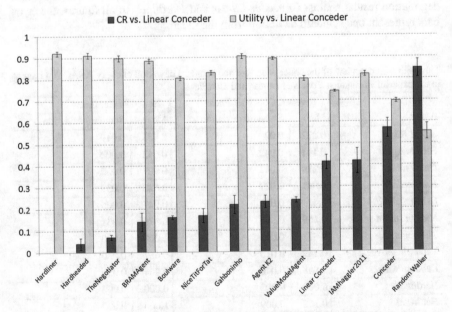

Fig. 8.6 Comparing the results versus the concession rate against the *Conceder Linear* opponent

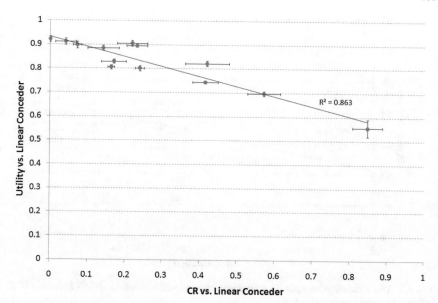

Fig. 8.7 The correlation between the concession rate and utility obtained against the *Conceder Linear* opponent of all ANAC 2011 and benchmark agents

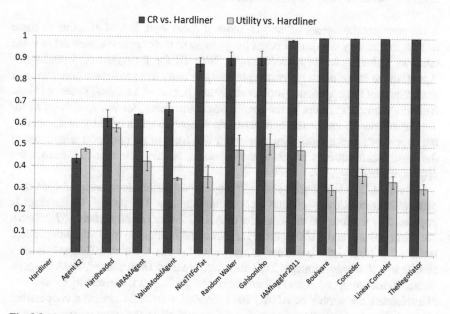

Fig. 8.8 Comparing the results versus the concession rate against the *Hardliner* opponent

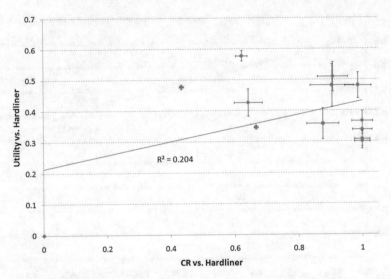

Fig. 8.9 The correlation between the concession rate and utility obtained against *Hardliner* of all ANAC 2011 and benchmark agents

8.4.3.1   Observations for ANAC 2011 Agents

If we consider the agents submitted, most of them were indeed adaptive to some degree, in that they take into account the strategy of their opponent. Several of these (such as *IAMHaggler2011, Gahboninho, Nice Tit for Tat Agent* or *ValueModelAgent*) make use of machine learning techniques to try and predict the concession or preferences of the opponent. Another important impact of a bidding strategy is how it *affects* subsequent offers by the negotiating counterpart. For example, *Nice Tit for Tat Agent* tries to entice the opponent into cooperative behavior by reciprocating the opponent's concessions. Such behavior is a form of second-level adaptivity, in the sense that it not only adapts to the opponent, but also attempts to influence the opponent's behavior.

All ANAC 2011 strategies reside in the left part of the chart, which means they play quite competitively against a conceding strategy. The top left of the chart is populated by the four 'nicer' strategies of ANAC 2011: *The Negotiator, Nice Tit for Tat, Gahboninho,* and *IAMhaggler2011.* Note that in addition to the time-dependent tactics, *The Negotiator* is located in the top of the chart, which indicates it is very similar to a *Boulware* strategy. The other four finalists: *HardHeaded, BRAMAgent, ValueModelAgent,* and *Agent K2* are more competitive. It is interesting to note that *HardHeaded,* the winner of ANAC 2011, concedes the least against a cooperating strategy. The fact that these strategies did very well in the competition seems to indicate that in order to be successful in such circumstances, an agent should behave competitively, especially against very cooperative strategies.

Now, it is interesting to see that, while the environment encourages flexibility, being adaptive does not necessarily benefit the agents in terms of winning the competition. In particular, it may come as a surprise that *HardHeaded*, a relatively simple and inflexible strategy, relying heavily on hard-coded parameters, achieved first place in ANAC 2011. There are several reasons for this. First of all, such a strategy relies on the opponent strategies to eventually adapt to its demands. Thus, if there are sufficient adaptive strategies, being competitive is a good strategy to use.

Indeed, it seems reasonable to conjecture that the nicer we play against a conceding opponent (meaning the higher our CR is against a conceding opponent), the less utility we make in the negotiation. To test this hypothesis, we selected *Conceder Linear* as the opponent, and then compared the CR of every agent against the average utility of the agreements being reached. The results are plotted in Fig. 8.6. It is readily observable that against this type of opponent, it pays to concede as little as possible. For example, *HardHeaded* has the lowest CR after *Hardliner*, which results in a very high score against *Conceder Linear*. In general, the correlation is very high (coefficient of determination of $R^2 = 0.86$) between the CR and utility obtained against *Conceder Linear* (see Fig. 8.9). This means that CR is a good predictor of how an agent fares against a cooperative agent. As noted, the CR is a measure of the bidding strategy. Therefore, against such a cooperative opponent, the decisive factors in the success of an agent are the circumstances and timing under which the bidding strategy makes its concessions.

In contrast, we cannot make similar statements about how to play against a very competitive agent such as *Hardliner*. As can be seen in Figs. 8.6 and 8.7, there is a positive correlation between the CR and the average utility obtained when playing *Hardliner*. This means that the nicer an agent plays, the more utility is obtained against *Hardliner*. This makes sense, as competitive play against *Hardliner* is sure to result in a break off. However, the correlation is very weak, with a coefficient of determination of $R^2 = 0.20$. Therefore in this case the CR is not a good predictor of the negotiation success. Because a break off of the negotiation is more likely against a very competitive opponent, the acceptance conditions of an agent play a much more important role than in the previous case.

Clearly, the only way to receive any utility in a negotiation against *Hardliner* is for an agent to give in fully. But in a tournament setting, an agent should not only take into account its own utility, but it should also make sure not to give away too much utility to the other contestants. Indeed, it is not entirely clear how one should successfully negotiate against the *Hardliner* (or similar strategies such as *HardHeaded*) without getting exploited in terms of tournament score. As a consequence, there exists a tension between the short-term negotiation incentives of a negotiating agent and its goal to get a high overall tournament ranking. Such considerations are related to the composition of the tournament pool [1], which we cover more deeply in Sect. 11.3.6 of our overall conclusions.

8.5   Conclusion and Discussion

Making concessions during a negotiation is vital for getting an agreement. Successful negotiations are only possible when both parties employ an effective conceding strategy. Designing a good strategy of when and how much to concede is challenging, and that is why there are many current negotiation implementations that concede in very different ways.

In this chapter, we characterized negotiation strategies by how they concede throughout the negotiation, and we did so by using a quantitative measure called the concession rate of an agent. We first formally defined the notion of concession rate as a normalized measure of the largest concession that was made during the negotiation. This formalizes the concept of an agent's willingness to concede against different opponents.

We then presented an empirical method to effectively compute the concession rate of agents, and then applied our approach to a selection of well-known agents (including all participants of ANAC 2010 and 2011) in an experimental setting. For the first time, this gives insight into the strategy space of negotiation tactics employed in ANAC. We subsequently used our method to classify the agents into four categories of concession behavior.

In addition to classifying agent strategies, various conclusions can be drawn based on charting the experimental results. Indeed, there is a wide spread in concession rates of current agents, and our classification chart is a useful method to cluster strategy types. The concession rate diagram shows that similar strategies have similar concession characteristics, and makes it easy to understand the agent's main negotiation characteristics at a glance.

Some extreme agents are located in the extreme regions of the chart, while the stronger agents form a cluster in the competitive corner. The results indicate that in order to be successful in an automated negotiation competition, an agent should not concede much, especially not to very cooperative strategies.

In general, the correlation is very high between the concession rate and utility obtained against such a conceding strategy. This means that a low concession rate is a good predictor for high performance against a conceding agent. The same behavior would be inappropriate against a very competitive agent that does not concede at all during the negotiation. We have also demonstrated that, in general, the nicer an agent plays against this type of opponent, the more utility it obtains. This is because, against such a hard-headed opponent, there is a high chance the negotiation would break down if no concessions are made.

This then leads to a dilemma for the agents to be either a teacher (who tries to entice their opponent to adapt by employing a tough strategy) or a learner (who tries to adapt to maximize its own utility, given the behavior of the opponent), which lies at the heart of many bilateral negotiation problems. If the opponent is flexible and adapting to one's demands, there is little point in conceding. However, if the opponent is being strictly hard-headed (even appearing irrationally so), reaching some agreement is typically preferable to no agreement. The importance of learning strategies that try first to detect the adaptivity of the opponent (such as *Gahboninho*, which proved very robust in ANAC 2011) is an important insight which could be taken up in further research in bilateral negotiations beyond the competition.

This chapter is based on the following publications: [1, 2]

Tim Baarslag, Koen V. Hindriks, and Catholijn M. Jonker. Towards a quantitative concession-based classication method of negotiation strategies. In David Kinny, Jane Yung-jen Hsu, Guido Governatori, and Aditya K. Ghose, editors, *Agents in Principle, Agents in Practice*, volume 7047 of *Lecture Notes in Computer Science*, pages 143–158, Berlin, Heidelberg, 2011. Springer Berlin Heidelberg

Tim Baarslag, Katsuhide Fujita, Enrico H. Gerding, Koen V. Hindriks, Takayuki Ito, Nicholas R. Jennings, Catholijn M. Jonker, Sarit Kraus, Raz Lin, Valentin Robu, and Colin R. Williams. Evaluating practical negotiating agents: Results and analysis of the 2011 international competition. *Articial Intelligence*, 198:73–103, May 2013

References

1. Baarslag T, Fujita K, Gerding EH, Hindriks KV, Ito T, Jennings NR, Jonker CM, Kraus S, Lin R, Robu V, Williams CR (2013) Evaluating practical negotiating agents: Results and analysis of the 2011 international competition. Artif Intell 198:73–103
2. Baarslag T, Hindriks KV, Jonker CM (2011) Towards a quantitative concession-based classi-fication method of negotiation strategies. In: Kinny D, Yung-jen Hsu J, Governatori G, Ghose AK (eds) Agents in principle, agents in practice, lecture notes in computer science, vol 7047. Springer, Berlin, pp 143–158
3. Deutsch M, Coleman PT, Marcus EC (2000) The handbook of conflict resolution: theory and practice. In: Jossey-Bass, 1st edn
4. Faratin P, Sierra C, Jennings NR (1998) Negotiation decision functions for autonomous agents. Robot Auton Syst 24(3–4):159–182
5. Fatima SS, Wooldridge MJ, Jennings NR (2002) Optimal negotiation strategies for agents with incomplete information. In: Revised papers from the 8th international workshop on intelligent agents VIII, ATAL '01. Springer, London, pp 377–392
6. Gode DK, Sunder S (1993) Allocative efficiency in markets with zero intelligence (ZI) traders: market as a partial substitute for individual rationality. J Polit Econ 101(1):119–137
7. Lewicki RJ, Saunders DM, Barry B, Minton JW (2003) Essentials of negotiation. McGraw-Hill, Boston
8. Pruitt DG (1981) Negotiation behavior, Academic Press
9. Sierra C, Faratin P, Jennings NR (1997) A service-oriented negotiation model between autonomous agents. In: Boman M, van de Velde W (eds) Proceedings of the 8th European workshop on modelling autonomous agents in multi-agent world, MAAMAW-97, lecture notes in artificial intelligence, vol 1237. Springer, pp 17–35

Chapter 9
Optimal Non-adaptive Concession Strategies

Abstract In the previous chapter we presented an empirical method to classify agents into four broad categories of concession behavior, and we formulated guidelines on how agents should bid against each category. This chapter follows an alternative approach by devising *optimal concession strategies* against specific classes of acceptance strategies. Many time-based concession strategies have already been proposed, but they are typically heuristic in nature, and therefore, it is still unclear what is the right way to concede toward the opponent. We provide a theoretical model in which a bidder makes a series of offers to an acceptor with unknown preferences. We apply *sequential decision techniques* as we did in Chap. 5, but this time to find analytical solutions that optimize the expected utility of the *bidder*, given certain strategy sets of the opponent. We then compare our solution to state of the art concession techniques in a negotiation simulation. Our solutions turn out to significantly outperform current approaches in terms of obtained utility. Our results open the way for a new and general concession strategy that can be combined with various existing learning and accepting techniques to yield a fully-fledged negotiation strategy for the alternating offers setting.

9.1 Introduction

A key insight of negotiation research is that making concessions is crucial to conducting a successful negotiation. There are important reasons to make concessions during the negotiation[1]: it is often used to elicit cooperation from the other, in the hope that the other will reciprocate in kind. Second, it conveys information to the opponent, both about the negotiator's preferences and about the perceptions of the opponent. But most importantly, it is the time pressure of the negotiation itself (typically in the form a deadline or a perceived maximum number of bidding rounds) that operates as a force on the parties to concede[2]. An approaching deadline puts important pressure on the parties to reduce their aspirations, especially when the time pressure heightens, which is referred to as the "eleventh hour effect".

Given the paramount importance of time in bargaining, it is not surprising that many negotiating agents adjust their level of aspiration based on the time that is left in the negotiation. There is a clear rationale behind the design of such agents, given their

© Springer International Publishing Switzerland 2016
T. Baarslag, *Exploring the Strategy Space of Negotiating Agents*,
Springer Theses, DOI 10.1007/978-3-319-28243-5_9

aim to maximize the chance of reaching an agreement in a limited amount of time. For example, well-known time dependent tactics (see Sect. 2.3.3), such as *Boulware* and *Conceder*, are characterized by the fact that they consistently concede throughout the negotiation process as a function of time. The same kind of time-based concession curves can also be observed in practice in ANAC (see Appendix B). However, in the TDT's, as well as in some very effective agents, such *Agent K* (winner of ANAC 2010) and *HardHeaded* (winner of ANAC 2011), the specific concession curve is selected rather arbitrarily, and is not informed by any other insights; therefore, they make largely unfounded choices on how much to concede at each time interval.

Alternatively, behavior dependent tactics such as Tit for Tat (cf. Sect. 2.3.3) base their decision to make concessions on the actions of the other negotiating party. However, such *adaptive* approaches do not give us any information on how to concede based on time alone.

Work that presents optimal choices of how much to concede includes game theoretic work (e.g. [3]) and single-shot bargaining, also known as the *ultimatum game* [4]. However, these approaches usually assume a complete information setting, or a game where the deal is struck immediately, which we cannot apply to a typical concession-based negotiation. Furthermore, this type of work typically revolves around equilibrium strategies, which assumes full rationality on the part of both agents. We are more interested in optimal solutions for one negotiating party, playing against various classes of acceptance strategies.

This chapter aims to find out how time pressure alone, in the form of a deadline, should influence the concession behavior of a negotiator against specific opponent classes. To do so, we employ methods from sequential decision theory to devise negotiation strategies that make *optimal* concessions at each negotiation round. Finally, we show that an agent making these optimal concessions performs better than any other in our experimental setup.

We begin with an example in Sect. 9.2 that sets the stage for our time-based concession model in Sect. 9.3. We apply our methods to find optimal concessions against opponents that accept according to acceptance thresholds in Sects. 9.4 and 9.5. We subsequently compare the optimal bidding strategy with state of the art bidding strategies in a series of tests (Sect. 9.6). We conclude our chapter with a discussion of the contributions of this chapter, as well as its implications (Sect. 9.7).

9.2 An Example

The following example serves as an illustration of the basic insights that have motivated our approach.

Suppose agent B is negotiating the purchase of a house with a buyer, agent A. As in the rest of this chapter, we will only focus on one of the two parties; in this case B.

B has set the opening price at \$300,000, but is (secretly) willing to go down to \$250,000 if necessary. B has to strike a balance between the probability that A buys the house, and getting as much utility out of the agreement as possible (i.e., not conceding too much). Our main question is: what offers or concessions should B make towards A in order to perform optimally and maximize his own outcome? That is, what is the right way to lower the price of the house depending on the remaining time and the acceptance policy of A?

Let us consider the easiest case where B has only *one* offer to make to A, which A can either accept or reject. If B makes an offer $x \in [\$250,000; \$300,000]$ to A, then this will yield him the following utility:

$$U(x) = \begin{cases} \frac{x - \$250,000}{\$50,000}, & \text{if } A \text{ decides to accept,} \\ 0, & \text{when } A \text{ rejects the offer.} \end{cases}$$

Now suppose A is prepared to pay up to \$280,000 for the house. Of course, B does not know this, but instead presumes that A's *acceptance threshold* could be anywhere between \$200,000 and \$300,000 with equal probability. It follows that B believes the chance that A accepts decreases linearly in terms of the price offered. Then, B's strategy is simple: he should set the price to

$$\arg \max_x U(x) \cdot P(A \text{ accepts } x), \tag{9.1}$$

which is equal to

$$\arg \max_x \left(\frac{x - \$250,000}{\$50,000} \right) \cdot \left(1 - \frac{x - \$200,000}{\$100,000} \right). \tag{9.2}$$

We can readily compute that the maximum is reached for $x = \$275,000$, and so with only one offer to make, B should pick this price in order to optimize his utility. Note how B's offer falls exactly in between his reservation price and his opening price. Fortunately for both, the price x is also actually lower than the maximum A was willing to pay, and therefore, she will buy the house.

This simple case serves as a good intuition to proceed to the more general setting we consider below.

9.3 Making Non-adaptive Concessions

Our negotiation model builds upon our definitions in Sect. 2.2.2: two agents A and B exchange offers in turns on a negotiation domain Ω. A and B have utility functions $U_A : \Omega \rightarrow [0, 1]$ and $U_B : \Omega \rightarrow [0, 1]$ and reservation values $rv_A, rv_B \in [0, 1]$ respectively. The agents will only propose offers that they deem acceptable, namely bids with higher utility than their reservation value. Likewise, the reservation values acts as the lowest utility they are willing to accept.

The process is illustrated in Fig. 9.1. With n rounds remaining, B starts by making an offer $B_n \in \Omega$, which A can either respond to with an accept or a counteroffer.

Fig. 9.1 Sequential diagram
of the negotiation process. B
starts the negotiation with
bid B_n, and after every
proposal, the other responds
with a counteroffer or an
ACCEPT

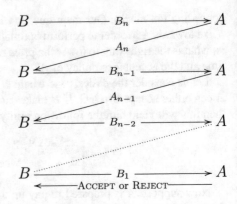

If A accepts, the negotiation ends in agreement, and both sides obtain their respective utility. If A rejects the offer, she replies with a counteroffer A_n, the number of remaining rounds decreases by one, and B can make a second offer, B_{n-1}. This process continues until B makes his last offer B_1. If A also turns down this last offer, A and B receive their respective reservation values rv_A and rv_B.

We are concerned with finding optimal concessions for one of the parties, given certain strategies used by the opponent. As in Chap. 5, we focus on agent B for his *Bidding* role, and on A for her *Accepting* role. In particular, we do not focus on what offers A should generate, or what acceptance policy B should employ.

There is an important class of strategies that allows us to make some further simplifications. In line with our aim to find concession curves that depend only on the time remaining, we will assume B's strategy is not adaptive (i.e., B does not change his behavior according to the bids that have been exchanged). This means B can completely ignore A's bids, and we can code them by ACCEPT or REJECT instead. Other examples of such agents include the family of time-dependent tactics, the *sit-and-wait* agent, and *IAMcrazyHaggler* (cf. Appendix C). Note that this then effectively defines an one-sided bidding protocol in which B submits n bids to A in a sequential manner, which is equivalent to a repeated ultimatum game with reservation values [5].

The fact that B does not adapt to his opponent puts him in a significantly more difficult epistemic position than in a typical negotiation, since B could normally gather valuable information from A's counter offers. This holds in a very fundamental sense: apart from the incoming offers, B can only distinguish n different possible states, namely the number of rejects that he received. This means that we can already compute the appropriate *concession curve* before the start of the negotiation; i.e., the bids or utilities that need to be sent out at every round.

Also note that as time moves *forward*, the indexing of the offers by B runs *backward*, just as in Chap. 5. This has the advantage that it simplifies the calculations of certain recurrence relations we encounter in this chapter, as the expected utility with $j + 1$ rounds remaining depends on the expected utility in the future (i.e., with j rounds remaining). Secondly, it allows us to define our model without a preset deadline, as we do not need to specify the maximum number of rounds beforehand.

9.4 Conceding and Accepting

The aim of this chapter is to find the right concession behavior for B given that he only has n offers to try out. B will only propose offers that he himself deems acceptable, namely bids in the following set:

$$O_B = \{\omega \in \Omega \mid U_B(\omega) \ge \mathrm{rv}_B\}. \tag{9.3}$$

B can make his bids in many ways. A well-known method using concession curves are the time-dependent tactics such as *Boulware* and *Conceder*, as defined in Sect. 2.3.3. When there are j rounds remaining (out of a total of n rounds), such strategies make a bid with utility closest to

$$P_{\min} + (P_{\max} - P_{\min}) \cdot \left(1 - k - (1 - k) \cdot \left(1 - \frac{j}{n}\right)^{\frac{1}{e}}\right),$$

for certain choices of k, e, $P_{\min}$ and $P_{\max}$, which control the concession rate and the minimum and maximum target utilities. These tactics form the basis of many successful negotiation strategies [6–9], in which variants of the above formula are used to decide the appropriate concessions, combined with advanced techniques such as preference modeling and strategy prediction. In this chapter, we will propose a different kind of concession curve that is also able to provide the groundwork for designing an advanced negotiation agent.

B's optimal strategy of course depends on how A chooses to accept or reject. While B makes his offers, A has to solve some kind of stopping problem to decide when the offer is sufficient.

Player A will have to either accept or reject a bid $\omega \in \Omega$ in every round j. A may update her beliefs in round j based on ω and the bidding history of the opponent, which is an element h_j from the set of offers $\mathcal{H}_j = \Omega^{n-j}$ made by B. Note that we only need to consider the bids made by B, since the rejects of A are already implicitly represented. For example, for $n = 8$ and $j = 5$, the information set of A is a history of rejected offers $h_5 = (B_6, B_7, B_8) \in \mathcal{H}_5$, adjoined with the bid B_5 of the current round.

Hence, we can represent A's acceptance strategy in round j as a function

$$\alpha : \mathcal{H}_j \times \Omega \to \{\text{ACCEPT}, \text{REJECT}\}. \tag{9.4}$$

We will focus on a specific set of strategies that accept according to *acceptance thresholds*. That is, A's acceptance strategy in round j is specified by a utility constant α_j (possibly dependent on rv_A) such that

$$\alpha(h_j, \omega) = \begin{cases} \text{ACCEPT} & \text{if } U_A(\omega) \ge \alpha_j, \\ \text{REJECT} & \text{otherwise.} \end{cases} \tag{9.5}$$

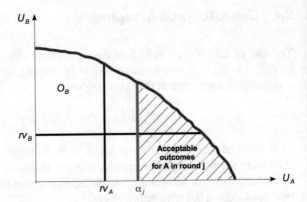

Fig. 9.2 A's acceptable offers for round j consist of all bids ω with $U_A(\omega) \geq \mathrm{rv}_A$, while B's possible offers O_B consist of all bids ω such that $U_B(\omega) \geq \mathrm{rv}_B$

We believe acceptance thresholds are a natural set of acceptance strategies to consider, since it is reasonable to assume that if A finds a bid acceptable, then so is any bid with higher utility.

One of the simplest acceptance strategies for A is *satisficing*: accepting any offer with a utility above her reservation value by setting all α_j equal to rv_A. This is done, for example, when a negotiator is more concerned with getting any deal at all than reaching the best possible deal (e.g., [10]). Needless to say, this is a very simple acceptance strategy, as A normally wants to get as much out of the negotiation as possible. In other applications, illustrated in Fig. 9.2, A's threshold would be higher at the beginning of the negotiation, and would slowly decrease towards rv_A. However, as we shall see in the next section, the case of a satisficing acceptor already requires highly effective conceding behavior by B.

A might also employ a more fundamental approach by trying to optimize her own utility, taking into account the number of rounds remaining. Optimal stopping theory provides optimal solutions for A for the case that A has incomplete information about B's offers. We repeat Proposition 5.1 from Chap. 5 p. 91, which expresses the optimal solution against a bidder B with completely unknown utility goals:

Proposition 9.1 *When B makes random bids of utility uniformly distributed in $[0, 1]$, and with j offers still to be observed, A's optimal acceptance strategy is to accept an offer of utility x exactly when $x \geq v_j$, where v_j satisfies the following equation:*

$$\begin{cases} v_0 = \mathrm{rv}_A, \\ v_j = \frac{1}{2} + \frac{1}{2}v_{j-1}^2. \end{cases} \qquad (9.6)$$

As it will turn out, this strategy is closely related to the strategy B should follow against satisficing acceptors.

9.5 Making Optimal Offers

We will now outline our general method to make optimal offers, which extends our running example from Sect. 9.2 to a general domain with an arbitrary number of remaining rounds.

Suppose A uses an acceptance strategy based on acceptance thresholds $(\alpha_0, \alpha_1, \ldots)$, with $\alpha_j \geq \mathrm{rv}_A$; that is, A accepts an offer $\omega \in \Omega$ with j remaining rounds exactly when $U_A(\omega) \geq \alpha_j$. Suppose we have $j + 1$ rounds to go, and B has to decide the optimal bid B_{j+1} to make.

The general formula for the expected utility $U_{j+1}(\omega)$ for B of offering $\omega \in O_B$ is as follows:

$$
U_{j+1}(\omega) = \begin{cases} U_B(\omega), & \text{if } A \text{ accepts,} \\ \text{the expected utility} \\ \text{with } j \text{ remaining rounds, if } A \text{ rejects.} \end{cases}
$$

When there are no more rounds remaining, B will get rv_B, hence $U_0(\omega) = \mathrm{rv}_B$. The recursive nature of this equation allows us to employ techniques from sequential decision theory to formulate the optimal way to make concessions.

We will write U_{j+1} for the highest expected utility B can obtain in round $j + 1$, which is thus given by the following equations:

$$
U_0 = \mathrm{rv}_B, \text{ and}
$$
$$
\begin{aligned}
U_{j+1} &= \max_{\omega \in O_B} U_B(\omega) \cdot P(U_A(\omega) \geq \alpha_{j+1}) \\
&\quad + U_j \cdot P(U_A(\omega) < \alpha_{j+1}) \\
&= U_j + \max_{\omega \in O_B} \left(U_B(\omega) - U_j \right) \cdot P(U_A(\omega) \geq \alpha_{j+1}).
\end{aligned}
$$

The corresponding optimal *bid* that B should make is given by the ω that maximizes the above equation, and therefore,

$$
B_{j+1} = \arg\max_{\omega \in O_B} \left(U_B(\omega) - U_j \right) \cdot P(U_A(\omega) \geq \alpha_{j+1}).
$$

Solving this equation would be straightforward, if it were not for the fact that B in general does not have full knowledge of a number of aspects in this equation: U_A is of course unknown to B, and so are the acceptance thresholds α_j. To make matters worse, the acceptance thresholds generally depend on the reservation value of A, which is also unknown to B.

B will therefore have to make some assumptions about U_A and her concession thresholds. Assume that B has an estimation of the reservation value of A, which does not change according to A's behavior. As in [11], the estimation is characterized by a probability distribution $F_j(x)$ for every remaining round j, where $F_j(x)$ denotes the probability that A's reservation value is no greater than x.

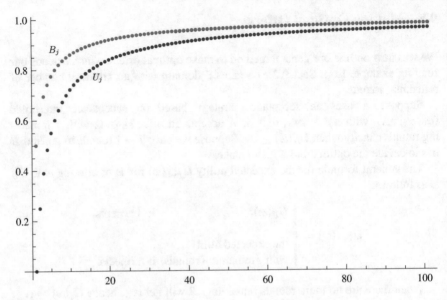

Fig. 9.3 Graph of B_j and U_j for remaining rounds $j \in \{0, 100\}$

We will now analytically solve the specific case of an opponent with a satisficing accepting strategy. To get concrete examples of concession curves (Fig. 9.3), we will consider the same classical buyer-seller scenario as in Chap. 5 to evaluate our solutions. We study the negotiation scenario of *Split the Pie* [3, 12], where two players have to reach an agreement $x \in [0, 1]$ on the partition of a pie of size 1. The pie will be partitioned only after the players reach an agreement. In this setting, we instantiate $\Omega = [0, 1]$ to represent a pie of size 1, with A and B having opposing preferences on them: $U_B(x) = x$ and $U_A(x) = 1 - x$.

We assume that A's acceptance strategy is limited to the class where she accepts any offer that is better than her reservation value; i.e., $\forall_j \alpha_j = \text{rv}_A$, where we assume rv_A is uniformly distributed. The probability that A accepts in round j is now:

$$P(U_A(\omega) \geq \alpha_j) = P(U_A(\omega) \geq \text{rv}_A) = F_j(U_A(\omega)). \tag{9.7}$$

The general formula now simplifies to

$$U_{j+1} = U_j + \max_{x \geq \text{rv}_B} (x - U_j) \cdot (1 - x). \tag{9.8}$$

Note that the following holds, even for the general setting:

Proposition 9.2 *When B has more rounds remaining, B can expect to get more utility out of the negotiation. That is, for every remaining round j, we have $U_{j+1} \geq U_j$.*

Proof B can always make the bid $\omega_{\max} = \arg\max_\omega U_B(\omega)$ with $j + 1$ rounds remaining, to get at least as much utility as in the next round.

With the help of Proposition 9.2, we can show the maximum in Eq. (9.8) is attained for $x = B_j$, where B_j satisfies the following relationship:

$$\begin{cases} B_1 = \frac{\text{rv}_B + 1}{2}, \\ B_{j+1} = \frac{1 + U_j}{2}. \end{cases} \tag{9.9}$$

This yields the following recurrence relation for U_j:

$$\begin{cases} U_0 = \text{rv}_B, \\ U_{j+1} = \frac{1}{4}\left(U_j + 1\right)^2. \end{cases} \tag{9.10}$$

With these equations, we can now compute the expected value of the optimal bids B has to make in terms of his own reservation value (see Fig. 9.4). For example:

$$U_1 = \frac{1}{4}\left(1 + \text{rv}_B\right)^2,$$

$$U_2 = \frac{1}{64}(5 + \text{rv}_B(2 + \text{rv}_B))^2,$$

$$\vdots$$

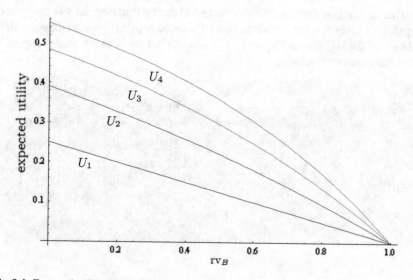

Fig. 9.4 Expected utility U_j for B with $j \in \{1, 2, 3, 4\}$ rounds to go, depending on B's reservation value rv_B

The corresponding optimal concessions by B are as follows:

$$B_1 = \frac{1 + rv_B}{2},$$

$$B_2 = \frac{1}{8} \cdot (5 + rv_B \cdot (rv_B + 2)),$$

$$\vdots$$

Note that both Eqs. (9.9) and (9.10) are expressed in terms of the value U_j. We obtain a much more elegant formulation when we express them in terms of B_j:

Proposition 9.3 *For all $j \geq 1$,*

$$B_{j+1} = \frac{1}{2} + \frac{1}{2} B_j^2,$$

and

$$U_j = B_j^2.$$

Proof Both statements follow from rewriting Eqs. (9.9) and (9.10).

This also means that Eq. (9.10) is related to the logistic map, as was the case in Eq. (5.4) of Chap. 5: if we substitute $U_j = 1 - 4x_j$ in Eq. (9.10) we get the equivalent relation of the logistic map $x_j = x_{j-1}(1 - x_{j-1})$ at $r = 1$, so we cannot expect to solve the recurrence relation.

Proposition 9.4 *For $rv_B = 0$, there exists the following connection between B_j, U_j, and the optimal stopping cut-off values v_j from Eq. (9.6):*

$$B_j = v_j,$$

and

$$U_j = v_j^2.$$

Proof When $rv_B = 0$, then $B_1 = \frac{1}{2}$, and consequently, the B_j sequence has the same starting value and definition as the optimal stopping sequence v_j.

Proposition 9.4 shows that optimal *bidding* against a satisficing acceptor with unknown reservation value is the mirrored version of Proposition 5.1 about optimal *accepting* against a bidder that makes unknown offers. In both cases, the idea is the same, but with switched roles: both the optimal bidder and the optimal stopper aim to pick the optimal utility threshold that simultaneously maximizes the expected utility of an agreement and the chance of acceptance, given stochastic behavior by the opponent. We conclude this section with an example that relates our results to the housing example of Sect. 9.2.

Example Our assumptions in this section are consistent with our housing example when we scale the pie Ω to [\$250,000; \$300,000], and when $rv_B = 0$ corresponds to the utility of selling the house for \$250,000.

Our results indicate that the seller of the house should act as follows: we see that indeed, $B_1 = \frac{1}{2}$, so with one bid remaining, B should make a bid halfway between \$250,000 and \$300,000, as we had calculated before, with an expected utility of $U_1 = \frac{1}{4}$.

Also, $B_2 = \frac{5}{8}$, which means with *two* bids remaining, B should offer

$$\$250,000 + \frac{5}{8} \cdot \$50,000 = \$281,250.$$

See Table 9.1 for other entries.

Note that since B does not learn anything about A, B actually overestimates the expected utility, and more so when the number of rounds increases (as can be seen in Table 9.1). This is because B is not able to deduce, from A's rejects, that high utility values for B are simply not attainable.

Table 9.1 The optimal offers to make and their expected utility in the housing example, given the remaining amount of bidding rounds

Remaining rounds	Optimal offer	Expected utility
1	\$275,000	0.25
2	\$281,250	0.39
3	\$284,766	0.48
4	\$287,086	0.55
5	\$288,754	0.60
10	\$293,055	0.74
100	\$299,060	0.96

9.6 Experiments

In order to test the efficacy of the optimal bidding technique given by Proposition 9.3, we integrated it into a fully functional negotiating agent. Given j remaining rounds it makes an offer with utility target B_j as defined by Eq. (9.9). It does not accept any offers and it does not model the opponent in any way.

For the opponents (side A), we selected various well-known negotiation agents available for our setting, including the top three agents of ANAC 2012, namely *CUHK Agent, AgentLG,* and *OMAC Agent* (see Appendix E). We also included the time dependent tactics *Boulware* (with concession factor $e = 0.2$), and *Conceder* ($e = 2$), and as a baseline, we included the *Random Walker* strategy. We then compared the optimal bidder's performance with the same set of strategies on side B. To analyze the performance of different agents, we employed GENIUS (Appendix A).

Note that some of the agents in our setup are originally designed to work with the alternating offers protocol, while we essentially employ a one-sided bidding protocol; however, since our model is a simplified alternating offers protocol, it is easy to adjust the agents to work in our setting: for the opponents, we ignore any bids that are sent out; i.e., when side A accepts, the negotiation ends in agreement, while A's counter-offers count as rejects and are ignored. In effect, this means only A's acceptance mechanisms are used.

For our negotiation scenario we use a discretized version of *Split the Pie*. We set $rv_B = 0$, and we selected varying reservation values for A: $rv_A \in \{0, 0.1, 0.2, \ldots, 0.9\}$. We ran our experiments using varying total number of rounds $n \in \{1, 20, 40, \ldots, 100\}$, and we repeated every negotiation session 5 times for statistical significance.

The results of our experiment are plotted in Fig. 9.5, with the total amount of rounds n varying between 1 and 100. As is evident from the results, the optimal

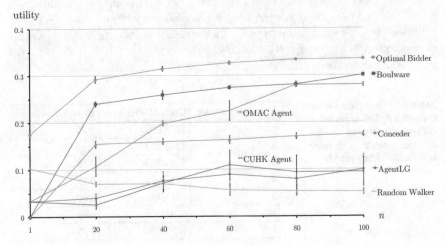

Fig. 9.5 The utility obtained by the bidding strategies in our experiments for different values of the total number of rounds n. The *vertical bars* indicate one standard deviation from the mean

bidder significantly outperforms all agents in all cases (*one-tailed t-test*, $p < 0.01$ for every n). The good relative performance of the optimal bidder is even more pronounced for $n = 1$, and it is easy to see why. Many of the other strategies will persist in aiming for a high utility, even with only few bids to send out. This results in many break-offs, while optimal bidder will settle for a much lower value in the last rounds with more potential agreements. For example, for $n = 1$, optimal bidder sets his utility target halfway between his reservation value and the maximum attainable; i.e., $B_1 = \frac{rv_B + 1}{2}$. On the other hand, with more bids remaining, optimal bidder acts as an extreme *Boulware* strategy, trying to get as much out of the negotiation as possible. Indeed, optimal bidder and *Boulware* tend to act more similar as n increases. *CUHK Agent* (the winner of ANAC 2012) and *AgentLG* obtain particularly low scores. The main reason for this is that these strategies are very behavior-dependent and do not take into account the remaining time as much as they need to.

Note that almost all agents obtain higher utilities with more negotiation rounds. This is to be expected, as more time allows for a more fine-grained search of what is acceptable for the opponent. The only exception is *Random Walker*, who only increases the chances to make disadvantageous bids for itself with more available time.

9.7 Conclusion

This chapter presents a theoretical model to calculate optimal concession curves against strategies that decide when to accept using acceptance thresholds. We compute these optimal concessions by employing sequential decision methods, and we show that they significantly outperform state of the art concession strategies, even against a much wider set of acceptance strategies. In specific instances, the optimal concession curve is equal in shape to the curve describing optimal stopping cut-off points. This confirms that optimal stopping theory is a powerful method that has many possible application in negotiation research. It is interesting to note that the concession curve that we found (as shown in Fig. 9.3) is not equal any time dependent curve (seen in Fig. 2.3 on page 27), but it bears most resemblance to a *Boulware*-like concession curve. As far as the authors are aware, this is the first time such informed time-based concession strategies have been formulated and tested in practice.

Our results demonstrate that our optimal bidding mechanism is an effective way for a negotiating agent to take into account the passing of time. We believe even more effective concession curves can be computed by studying broader ranges of acceptance strategies than we did in this chapter. Further improvements could be made by introducing a form of learning to the optimal bidder, using the BOA framework. Eventually, we envision a design of an automated negotiator that incorporates our optimal concession curve with regard to time-related concessions, while other types of concessions (e.g., to reduce costs [13], to elicit cooperation [14], or to convey information [15]) are handled separately by other concession modules.

This chapter is based on the following publications: [16]

Tim Baarslag, Rafik Hadfi, Koen V. Hindriks, Takayuki Ito, and Catholijn M. Jonker. *Optimal non-adaptive concession strategies with incomplete information*. In *Proceedings of The Seventh International Workshop on Agent-based Complex Automated Negotiations (ACAN 2014)*, 2014

References

1. Rubin JZ, Brown BR (1975) The social psychology of bargaining and negotiation. Academic press, New York
2. Carnevale PJD, Lawler EJ (1986) Time pressure and the development of integrative agreements in bilateral negotiations. J Conflict Resolut 30(4):636–659
3. Rubinstein A (1982) Perfect equilibrium in a bargaining model. Econometrica 50(1):97–109
4. Raiffa H (1982) The art and science of negotiation: How to resolve conflicts and get the best out of bargaining. Harvard University Press, Cambridge
5. Slembeck T (1999) Reputations and fairness in bargaining—experimental evidence from a repeated ultimatum game with fixed opponents. Experimental, EconWPA
6. Baarslag T, Fujita K, Gerding EH, Hindriks KV, Ito T, Jennings NR, Jonker CM, Kraus S, Lin R, Robu V, Williams CR (2013) Evaluating practical negotiating agents: results and analysis of the 2011 international competition. Artif Intell 198:73–103
7. Fatima SS, Wooldridge MJ, Jennings NR (2002) Optimal negotiation strategies for agents with incomplete information. In: Revised papers from the 8th international workshop on intelligent agents VIII, ATAL '01. Springer, London, pp 377–392
8. Kawaguchi S, Fujita K, Ito T (2012) Compromising strategy based on estimated maximum utility for automated negotiating agents. In: Ito T, Zhang M, Robu V, Fatima S, Matsuo T (eds) New trends in agent-based complex automated negotiations. Series of studies in computational intelligence. Springer, Berlin, pp 137–144
9. van Krimpen T, Looije D, Hajizadeh S (2013) Hardheaded. In: Ito T, Zhang M, Robu V, Matsuo T (eds) Complex automated negotiations: theories, models, and software competitions. Studies in computational intelligence, vol 435. Springer, Berlin, pp 223–227
10. Niemann C, Lang F (2009) Assess your opponent: a bayesian process for preference observation in multi-attribute negotiations. In: Ito T, Zhang M, Robu V, Fatima S, Matsuo T (eds) Advances in agent-based complex automated negotiations. Studies in computational intelligence, vol 233. Springer, Berlin, pp 119–137
11. Li C, Giampapa J, Sycara KP (2006) Bilateral negotiation decisions with uncertain dynamic outside options. IEEE Trans Syst Man Cybern Part C Appl Rev 36(1):31–44
12. Stahl I (1972) Bargaining theory. Economic Research Institute, Stockholm
13. Baarslag T, Gerding EH, Aydoğan R, Schraefel MC (2015) Optimal negotiation decision functions in time-sensitive domains. In: 2015 IEEE/WIC/ACM international joint conferences on Web Intelligence (WI) and Intelligent Agent Technologies (IAT)
14. Baarslag T, Hindriks KV, Jonker CM (2013) A tit for tat negotiation strategy for real-time bilateral negotiations. In: Ito T, Zhang M, Robu V, Matsuo T (eds) Complex automated negotiations: theories, models, and software competitions. Studies in computational intelligence, vol 435. Springer, Berlin, pp 229–233
15. Baarslag T, Hindriks KV (2013) Accepting optimally in automated negotiation with incomplete information. In: Proceedings of the 2013 international conference on autonomous agents and multi-agent systems, AAMAS'13. International Foundation for Autonomous Agents and Multiagent Systems, Richland, SC, pp 715–722
16. Baarslag T, Hadfi R, Hindriks KV, Ito T, Jonker CM (2014) Optimal non-adaptive concession strategies with incomplete information In: Proceedings of the seventh international workshop on agent-based complex automated negotiations (ACAN 2014)

Chapter 10
Putting the Pieces Together

Abstract In the previous chapters, we performed extensive research into three different aspects of an automated negotiator: *bidding* (Chaps. 8 and 9), *learning* (Chaps. 6
and 7), and *accepting* (Chaps. 4 and 5). We have found novel ways to handle each
component, sometimes even in optimal ways for particular circumstances. The natural question then arises: *how do the pieces fit together?* Or, more specifically: *how
do the components of a negotiating agent influence its overall performance, and
which components are the most important for the end result of an agent?* In this
chapter, we provide an answer to this question in a quantitative way. In doing so,
we show that the BOA framework not only provides a useful basis for developing
and evaluating agent components, but also provides a powerful agent design tool.
Furthermore, we demonstrate that combining effective key components from different agents improves an agent's overall performance. This validates the analytical
approach of the BOA framework towards optimizing the individual components of
a negotiating agent. By combining agent components in varying ways, we are able
to demonstrate the contribution of each component to the overall negotiation result,
and thus determine the key contributing components. Moreover, we study the interaction between components and present detailed interaction effects. We find that the
bidding strategy in particular is of critical importance to the negotiator's success
and far exceeds the importance of opponent preference modeling techniques. Our
results contribute to the shaping of a research agenda for negotiating agent design by
providing guidelines on how agent developers can spend their time most effectively.

10.1 Introduction

Throughout this thesis, we performed extensive research into three different aspects
of an automated negotiator. One key components is the *bidding strategy*; focusing
on what kind of offers the negotiation strategy should choose, and in particular, what
kind of concessions should be made (see Chaps. 8 and 9). We also put emphasis
on *learning* techniques to model various opponent attributes, such as the (partial)
preference profile or the opponent's next move (Chaps. 6 and 7). Finally, there is the

© Springer International Publishing Switzerland 2016   181
T. Baarslag, *Exploring the Strategy Space of Negotiating Agents*,
Springer Theses, DOI 10.1007/978-3-319-28243-5_10

question of when to *accept*; i.e., under which circumstances an offer by the opponent should be agreed to (Chaps. 4 and 5).

In the preceding chapters, we found effective components for each of these aspects by focusing on one particular component. However, the interactions and relative importance of the individual components have not been studied in detail before. Some components may have a stronger impact on the performance than others, and there could be strong interdependencies between the components (e.g., a very competitive strategy hampering a learning technique by not exploring enough options). Thus, the main question of this chapter is: *How do the components of a negotiating agent influence its overall performance, and which components are the most important?*

Answering this question requires us to bring together so-far unconnected research on various elements of negotiating agent design, and to research whether combining effective key components from different agents improves an agent's overall performance. In this chapter, we show that the BOA framework not only provides a useful basis for developing and evaluating agent components, but also provides a powerful tool for designing full negotiating agents. We validate the analytical approach of the BOA framework by demonstrating that combining effective key components from different agents also improves an agent's *overall* performance.

We investigate the importance and relations between three key components of the BOA framework: the bidding strategy, the opponent model and the acceptance strategy. The question then becomes: *what is more important for a negotiation strategy to do well: to bid, to learn about the opponent, or to accept?* Or, more formally, how do each of the three components contribute to the effectiveness of a negotiation agent, and what are the interaction effects between them? We make these questions precise by formulating them in quantifiable terms of predictability of variance, and we determine the contribution of each component using the statistical measure of effect size. Once the individual contribution of each key component is established, we focus on the effects of combining components.

This process will give us a better understanding of how we can improve negotiation agents, as well as provide guidelines for the design of negotiating agents. Our findings indicate that the bidding strategy is by far the most important component, while the significance of learning about the opponent's preferences is rather small in comparison. Given the current focus on opponent learning techniques in automated negotiation research, we argue that more effort needs to be made to formulate effective bidding strategies.

We start by outlining our method to quantify importance of agent components in Sect. 10.2, with the notion of effect size. Section 10.3 presents our experiments, in which we combine a large set of agent components, followed by Sect. 10.4, which studies the contribution and interaction effects of each component. Finally, Sect. 10.5 presents our conclusions and recommendations for a research agenda on agent design.

10.2 Measuring the Contribution of Strategy Components

To determine the importance of a particular agent component in relation to another, and to measure the interactions between them, we use the notion of *effect size*. The effect size is a statistical measure to quantify the interactions between various controlled variables and measures their effect on the dependent variable, which in our case is the average outcome utility of the agent. In this way, the effect size expresses *how* significant or important a variable is [1]. This goes beyond the question of *whether* a specific variable is significant or important, which can be answered with standard statistical significance testing. We use the relative proportion of variation, known as the η^2 measure.

The η^2 measure examines the variance explained by every component to determine its importance. When the contribution of a component is small (a low η^2 value), there is little variation and the difference between the different components is small. This makes the choice of which component to use less important. On the other hand, if the contribution of a component is large (a high η^2 value), then the choice of a component is sure to have a big impact.

The η^2 measure rates the variance that a variable introduces using the *Sum of Squares*. The *Total Sum of Squares* (SS_{Total}) expresses the total dispersion of all data points; i.e., the total variance found within the data, which can be calculated from the differences between the data points x_i and the total mean $\bar{x}$ as follows:

$$SS_{Total} = \sum_{i=1}^{n}(x_i - \bar{x})^2. \tag{10.1}$$

The *Sum of Squares Between Groups* (SS_b) determines the variance between the different *groups* and is calculated as follows:

$$SS_b = \sum_{j=1}^{|G|} n_j(\bar{x}_j - \bar{x})^2, \tag{10.2}$$

where the $\bar{x}_j$ represents the group mean, $|G|$ represents the number of groups and n_j is the number of data points in group j.

To calculate the η^2 value for each variable i, we use Eq. (10.3), where SS_{b^i} is the sum of squares between groups for variable i:

$$\eta_i^2 = \frac{SS_{b^i}}{SS_{Total}} \tag{10.3}$$

A benefit of using this measure is that it allows for easy comparison between the different η^2 values on an interval scale, because the η^2 values are scaled according to the total variation (SS_{Total}). This is not the case with other effect size measures (e.g. partial η^2), which only allows for an ordinal scale comparison [2]. Another advantage of using η^2 is that it can be expressed in the form of percentages, making it more intuitive to understand. For example, if a certain controlled variable v has a η^2 value of 0.35, it means that 35 % of the variation in the dependent variable can be accounted for (or explained by) v.

10.3 Experiments

To analyze the relative importance and interactions of each strategy component, we require a wide array of tactics and techniques for every component. To explore the space of BOA components, we need a representative selection for every component from the BOA framework repository, ranging from baseline techniques to state of the art techniques. For every component, we aimed to select many variants, given that they were designed for a negotiation setting consistent with ours, with publicly available code, and consistent with the BOA framework (i.e., generic enough so that we could freely interchange them with arbitrary other components). To ensure as much variety as possible, we included components not only from this thesis, but also components designed by various researchers.

We employ the same negotiation setting of a bilateral negotiations with the alternating-offers protocol as in our definitions in Chap. 2. To make the extensive experiments feasible, we used a discrete time line as in Chap. 7, with a specified number of rounds $N = 3000$. Both agents receive utility 0 if they do not succeed in reaching an agreement before the deadline.

Our range of components are shown in Table 10.2. We selected 11 different bidding strategies in total. We selected state of the art bidding strategies from the top three strategies of the ANAC 2011 and 2012 competitions (see Appendices D and E), and as baseline bidding tactics, we included the time-dependent tactics *Boulware* (with concession rate $e = 0.2$), *Conceder Linear* ($e = 1$), *Conceder* ($e = 2$), and behavior-dependent strategies such as *Nice Tit for Tat*, *Absolute Tit for Tat* and *Relative Tit for Tat* (all defined in Sect. 2.3.3).

All bidding strategies were combined with six different opponent models. We selected five state of the art opponent models from ANAC that we could freely combine with arbitrary bidding strategies. To ensure a fair representation of opponent models, we selected from two main categories as defined in Chap. 6; *Bayesian models*, which use Bayesian learning techniques to create and update a model of the opponent's preference profile [3] and *Frequency models*, which keep track of how often certain items are requested by the opponent. As a baseline, we chose *No Model*, which selects bids for the opponent at random.

For the state of the art acceptance conditions, we selected a number of sophisticated strategies from top ANAC agents. We also included the most advanced version of the optimal stopping acceptance policy from Chap. 5 (*Optimal Stopping GPR*), which predicts the opponent's strategy using a Gaussian process regression technique to determine the probability that a better bid will be offered in the future. Lastly, we added the best performing acceptance strategies from Chap. 4, namely $\mathbf{AC}_{combi}(T, \mathrm{MAX}^W)$ and $\mathbf{AC}_{combi}(T, \mathrm{AVG}^W)$, with $T = 0.95$. We selected a number

Table 10.1 Characteristics of the negotiation scenarios

Scenario name	Size	Bid distrib.	Opposition
Car	15625 (*med.*)	0.136 (*low*)	0.095 (*low*)
Grocery	1600 (*med.*)	0.492 (*high*)	0.191 (*med.*)
Itex versus Cypress	180 (*low*)	0.222 (*med.*)	0.431 (*high*)
Laptop	27 (*low*)	0.295 (*med.*)	0.178 (*med.*)
Travel	188160 (*high*)	0.416 (*high*)	0.230 (*med.*)

of simple baseline acceptance policies as defined in Chap. 4, such as: $AC_{const}(c)$ ($c \in \{0.6, 0.7, 0.8, 0.9\}$), which accepts exactly when the utility of the opponent's offer is higher than a constant threshold c; $AC_{gap}(g)$, which accepts when the utility gap between the parties is smaller than a gap size $g \in \{0.1, 0.2\}$; and $AC_{next}(\alpha, \beta)$ ($\alpha \in \{1.0, 1.1, 1.2\}$, $\beta \in \{0, 0.1, 0.2\}$), which accepts when $\alpha \cdot u' + \beta \geq u$, where u is the utility of the bid that is ready to be sent out and u' is the utility of the opponent's offer.

For our opponent pool, we selected a representative set of 7 agents, ranging from baseline strategies (*Boulware* and *Conceder Linear*) to some of the top performing agents from the ANAC competitions (the top 2 agents from ANAC 2011: *HardHeaded* and *Gahboninho* [4], and number 1 and 3 from ANAC 2012: *CUHKAgent* [5] and *The Negotiator Reloaded*). Note that we did not run all agents against each other, but elected a representative opponent set instead. A full tournament with all negotiation agents would be unfeasible, since much smaller tournaments (such as ANAC) already take weeks to complete.

The negotiation scenarios were chosen on the basis of the following characteristics: *domain size* (number of possible bids), *bid distribution* (average distance of all bids to the nearest Pareto-optimal bid), and the *opposition of the domain* (the distance from the Kalai-Smorodinsky point to complete satisfaction). We picked 5 negotiation scenarios used in ANAC such that every characteristic varies between *high*, *medium* and *low*. The scenario information is summarized in Table 10.1 (we refer to Tables C.3 and D.3 for more details).

We created a large number of negotiation agents by combining all BOA components into full negotiation agents; i.e., we created a pool of 11 bidding strategies × 6 opponent models × 24 acceptance conditions, which amounts to 1584 negotiation agents in total.[1] We let all of them play against the 7 opponents in each scenario listed in Table 10.1, keeping track of their obtained utility in every negotiation session. To run our tournament we used the BOA framework as implemented in GENIUS (Appendix A). Since not all the negotiation strategies are deterministic, the tournament setup was run 5 times in order to reduce the amount of variance in the data, resulting in 277200 negotiation sessions in total (Table 10.2).

---

[1]Note that this set also includes already existing agents such as *HardHeaded* and *The Negotiator Reloaded*, since their components occur in all three groups.

Table 10.2 All negotiation strategy components used in experimental setup

Bidding, opponent modeling, and accepting components

Bidding strategy	Acceptance conditions
AgentLG	$AC_{combi}(T, AVG^W)$
CUHK agent [5]	$AC_{combi}(T, MAX^W)$
HardHeaded [4]	$AC_{const}(c)$
IAMhaggler2012	$AC_{gap}(g)$
The negotiator reloaded [6]	$AC_{next}(\alpha, \beta)$
Absolute tit for tat [7]	Agent K2 [4]
Nice tit for tat [8]	AgentLG
Relative tit for tat [7]	AgentMR [9]
Time-dependent tactics [7]	BRAMAgent2 [4]
	HardHeaded [10]
Opponent model	IAMhaggler2012
Agent Smith (Frequency) [11]	Nice Tit For Tat [8]
HardHeaded (Frequency [10]	OMAC Agent [12]
IAMhaggler (Bayesian) [13]	Optimal Stopping GPR
NASH Agent (Frequency)	The Negotiator [14]
The negotiator reloaded (Bayesian) [6]	The Negotiator Reloaded [6]
No model	

10.4   Component Contribution

To determine the contribution of each BOA component, we calculated the η^2 of the three components, using as input the utilities of all component combinations averaged over all the runs, domains and opponents. The calculated η^2 values are presented in Table 10.3 and visualized in Fig. 10.1.

Table 10.3 The η^2 measure for every component and the standard deviation of η^2 over the different runs, domains and opponents

Component	η^2	Standard deviation		
		Runs	Domains	Opponents
Bidding strategy (BS)	0.582	0.003	0.118	0.163
Opponent model (OM)	0.035	0.002	0.020	0.037
Acceptance conditions (AC)	0.118	0.003	0.121	0.071
BS * AC	0.114	0.003	0.023	0.082
OM * AC	0.014	0.001	0.011	0.016
BS * OM	0.085	0.004	0.040	0.040
BS * OM * AC	0.051	0.002	0.037	0.051

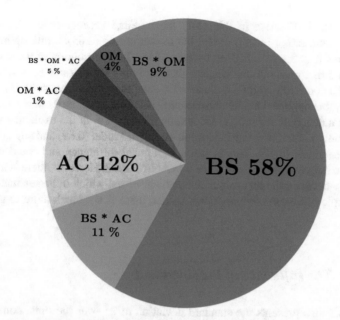

Fig. 10.1 Visual representation of the contribution of components

From the calculated contributions, we can observe that the bidding strategy is by far the most important component, accounting for 58 % of the variation in the negotiation strategy's performance, which is significantly higher (*one-tailed t-test,* $p < 0.01$) than the other components. Recall that the bidding strategy controls the concession speed of the agent, thereby managing the rate with which it moves towards the opponent according to its own utility. This is expected to have a huge impact on the final outcome, as it determines the subset of the outcome space an agreement can possibly be made in. We can also observe that the variance of the bidding strategy contribution is rather high, especially with respect to the opponent pool; this is explored further in Sect. 10.4.1.

Note that our method allows us to express the effectiveness of a specific choice for any given strategy component. For example, we can fix the bidding strategy to *CUHK Agent,* and calculate its average obtained utility when combined with all opponent models and acceptance conditions and then compare this with the average utility of another choice of the bidding strategy, for example *Conceder.* In fact, the difference in this case is particularly large, because using this method, *CUHK Agent* (bidding strategy of the winner of ANAC 2012) obtains the highest score of 0.79, while *Conceder* (a baseline bidding strategy) has the lowest score of 0.63. Such a utility difference means the difference between place 1 and place 8 in ANAC 2012 (see Table E.2), which gives a good indication of how important the bidding strategy is.

The second most important component is the acceptance condition, with a contribution of 12 %, which is significantly higher than the opponent model (*one-tailed t-test*, $p < 0.01$). Also in this case, the baseline acceptance conditions (e.g., $\mathbf{AC}_{const}(c)$ and $\mathbf{AC}_{next}(\alpha, \beta)$) obtain the lowest scores (0.65), while optimal stopping from Chap. 5 scores highest by far (0.75). The contribution of the acceptance condition still comes out low because these are only the extreme scores, and most acceptance condition scores are clustered around the group average (0.71).

Finally, the opponent model makes a surprisingly small contribution to the performance of a negotiation strategy, accounting for only 4 % of the explained variance. There is only a small difference between using *No Model* (0.69) and any of the state of the art models (0.72). This is still a significant difference and could make the difference between place 1 and place 5 in ANAC 2012. However, there is almost no difference between the learning methods themselves, which indicates that once an agent employs a reasonable opponent model, there is not much more to gain after that.

10.4.1 The Influence of the Opponent

Table 10.3 also presents the standard deviation of η^2 over the runs, domains and opponents for each of the components. This gives an indication of how much an additional run, domain or opponent would affect the average contribution of each component. Adding more runs will have little effect on the components' contribution, but the same cannot be said for the domains and opponents, which both have high standard deviations, especially for the bidding strategy and acceptance condition, which indicates that these components highly interact with the domain and opponent.

As we can see from Table 10.3, the opponent is the most important source of variance when considering component contribution. We take a closer look at this phenomenon by focusing on the subset of opponents that employ time-dependent tactics, comparing the η^2 values when negotiating against each of them. The advantage of focusing on the time-dependent opponents is that they are from the same family of tactics, so that only one factor is altered between the different opponents, namely the rate at which they concede. To get a more complete picture, we add two more time-dependent tactics to our opponent pool, namely *Extreme Boulware* ($e = 0.02$) and *Extreme Conceder* ($e = 5$).

The η^2 values are listed in Table 10.4 and presented in Fig. 10.2. They show a clear trend for the contribution of the strategy components. As the value of e increases (raising the cooperativeness of the opponent) the contribution of the bidding strategy decreases dramatically, from 77 to 42 %. Through closer inspection of the negotiation dynamics between the agents, we are able to explain this trend.

When the opponent concedes very little, it is up to the bidding strategy of the agent to yield to the opponent to avoid the consequences of no agreement. This places a lot of importance on the bidding strategy, because the speed at which the agent concedes dictates the agreement that will be reached, hence its large η^2 value. In these situations, the importance of having an effective opponent model is at its

Table 10.4 η^2 playing against time dependent opponents

Component	η^2 against opponents			
	$e = 0.02$	$e = 1$	$e = 2$	$e = 5$
BS	0.770	0.675	0.600	0.419
OM	0.010	0.014	0.007	0.004
AC	0.053	0.163	0.234	0.446
BS * AC	0.077	0.076	0.083	0.073
OM * AC	0.005	0.002	0.003	0.002
BS * OM	0.063	0.059	0.056	0.035
BS * OM * AC	0.022	0.013	0.018	0.020

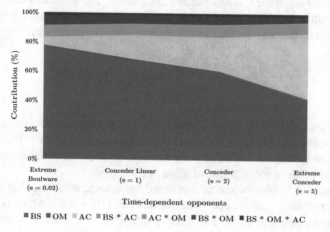

Fig. 10.2 The η^2 values of all components against different time-dependent opponents

peak, since it is both more challenging to achieve an acceptable outcome and it is harder to learn the opponent's preferences. Also, when an opponent makes very small concessions, the offers differ very little in utility. This makes the role of the acceptance condition less important, because the moment the agreement is reached is not as significant.

On the other hand, when the opponent is very cooperative, the choice between accepting and waiting for more concessions can make a big difference in the achieved outcome. Therefore, the acceptance condition of the agent and the bidding strategy of the opponent will primarily dictate the meeting point between the two agents. This is why, as the opponent concedes more, the contribution of the acceptance condition increases from 5 to 47 % at the cost of the bidding strategy.

10.4.2   Interaction Effects

Agent components do not perform their function in isolation. First of all, they interact with each other: for example, a learning method may be far less effective when the bidding strategy does not select enough 'exploratory offers' to learn more about the opponent's preferences. There are also interaction effects with the environment, such as the negotiation domain (e.g., learning about the opponent may be harder in big contract spaces), and the opponent (e.g., a good acceptance strategy is more important when the opponent is likely to make attractive offers).

The presence of these interactions requires a more thorough analysis, because the impact of one component can depend on the level of another. We denote interaction effects as $C_1 * C_2$, where C_i represents a component. The term *interaction effect* is used to quantify the effects of the interactions between the components. To determine the η^2 for an interaction we need to calculate the $SS_{b^{c_1*c_2}}$ as follows:

$$SS_{b^{c_1*c_2}} = \sum_{i=1}^{|C_1|} \sum_{j=1}^{|C_2|} (x_{ij} - \bar{x}_i - \bar{x}_j + \bar{x})^2 \tag{10.4}$$

This is then divided by SS_{Total} to determine the η^2 value for the interaction. The component interaction effects are listed at the bottom of Table 10.3. We will cover two types of interaction effects: the interactions between the components, and the interactions between the negotiation agent and its opponent.

There are two important interactions that deserve some attention. One is between the bidding strategy and the acceptance conditions ($BS * AC$), which is the highest interaction effect of 11 %. The bidding strategy and acceptance condition can be viewed as two simultaneous processes that can each result in a potential agreement. One process consists of offering bids that are appealing to the opponent in the hope that it will be accepted (bidding strategy), the other consists of receiving offers from the opponent and deciding whether these should be accepted or not (acceptance condition). These two components complement each other, as a good acceptance condition can compensate for a bad performing bidding strategy by accepting bids from the opponent, while a good bidding strategy can compensate for a bad acceptance condition by offering enticing counter bids.

Another important interaction is between the bidding strategy and the opponent model ($BS * OM$), which, with 9 %, is the second highest interaction effect. The opponent model directly influences the bidding strategy by aiding it in offering bids that are appealing to the opponent, thereby improving the chances of an agreement. Conversely, the effectiveness of an opponent model depends on how it is employed. Should the bidding strategy not use the opponent model to its full potential, then its effectiveness will be diminished. For example, the bidding strategy from *BRAM Agent* presents the opponent model with only a small selection of bids, combining only the ten most recent offers of the opponent, which reduces the effectiveness of the opponent model.

Interaction also exists between the acceptance conditions and the opponent model ($AC * OM$); however, this contribution is rather small (1 %). This is because the purpose of the opponent model is to model the opponent's attributes, while the acceptance condition typically acts according to the agent's own utility.

There are also three way interactions involving all three strategy components ($BS * OM * AC$), which account for 5 % of the variance (calculated analogously to Eq. (10.4)). For example, there are agents that not only use the opponent model to determine the best bid to offer to the opponent, but also use to determine their target utility. More sophisticated acceptance conditions make use of this target utility and are thus also affected by the opponent model, causing a three way interaction.

10.4.3 Combining the Best Components

Given the interaction effects between the different components, it is not immediately clear whether combining the best components together result in the most effective negotiation strategies. Note that we combined $11 \times 6 \times 24$ components, resulting in large number (1584) of agents. We could simply test whether the highest scoring agent out of the 1584 is indeed composed of the the best B, O, and A component, but we elected to take a more general approach. We first categorize the components in groups, and then test whether the top components together can assemble the top agents resulting from all group combinations.

We used the *Jenks Natural Break Algorithm* [15] for the classification of the individual components. *Jenks Natural Break Algorithm* is a data classification method designed to separate values into different groups. The algorithm attempts to find a series of natural break or separation values, which helps cluster the data in a natural way instead of an arbitrary classificatory scheme (e.g. dividing the space into equal intervals). The algorithm searches for the best break values such that it minimizes the variance within the groups and maximizes the variance between the groups.

We used the *Jenks Natural Break Algorithm* to divide each BOA component listed in Table 10.2 into three different performance groups (*High*, *Medium*, and *Low*), based on their average utility score. In order to determine whether combining better performing components would result in effective negotiation strategies, we combined these groups into 27 agent sets, of which we then tested the average utility (see Table 10.5).

Our results show that the agent set that employed the best performing components in isolation (i.e., the agents that were comprised of components that were all in the *High* category) have the highest average utility, performing significantly (*one-tailed t-test*, $p < 0.01$) better than all other agent groups. This means that indeed, the top components together produce the top agents. This shows that independently optimizing the individual components of the BOA framework is a feasible approach for developing negotiating agents. The agent group that used the worst performing components (*Low*, *Low*, *Low*) has the lowest average utility. The agent group (*Medium*, *Medium*, *Medium*) is found somewhere in the middle of the rankings, as expected.

Table 10.5 Rankings of 27 agent sets that were created by combining all BOA components that were each categorized to have *High*, *Medium* or *Low* individual performance

Component combinations			Avg. Util.	Std Dev
BS	OM	AC		
High	High	High	0.790	0.001
High	Medium	High	0.761	0.003
High	Low	High	0.761	0.003
High	High	Medium	0.751	0.001
High	Medium	Medium	0.739	0.001
Medium	High	High	0.739	0.001
High	Low	Medium	0.737	0.001
Medium	Medium	High	0.726	0.003
High	High	Low	0.726	0.001
Medium	High	Medium	0.713	0.001
High	Medium	Low	0.713	0.002
High	Low	Low	0.708	0.003
Medium	Medium	Medium	0.706	0.003
Medium	High	Low	0.684	0.001
Medium	Low	High	0.678	0.001
Low	Medium	High	0.678	0.002
Medium	Low	Medium	0.669	0.001
Low	Medium	Medium	0.669	0.001
Low	High	High	0.669	0.001
Medium	Medium	Low	0.666	0.002
Low	High	Medium	0.656	0.001
Low	Low	Medium	0.639	0.002
Low	Medium	Low	0.631	0.002
Low	High	Low	0.628	0.002
Low	Low	High	0.625	0.002
Medium	Low	Low	0.609	0.001
Low	Low	Low	0.608	0.001

10.5 Conclusion

This chapter investigates the performance effects of combining different instantiations of key components of existing negotiating agents, namely the bidding strategy, the opponent model, and the acceptance conditions. For this purpose, we analyzed the key components of a large set of both baseline and state of the art agents. We analyzed each component independently as well as in combination with the other components, using the measure η^2 to quantify their effect.

We found that combining the best agent components indeed results in the strongest agents. This shows that the three-component view of the BOA architecture not only provides a useful tool for developing negotiating agents, but this also validates the analytical approach of optimizing its individual components. By varying the key components of automated negotiators, we are able to demonstrate the contribution of each component to the negotiation result, and thus analyze the significance of each. Moreover, we are able to study the interaction effects between them. With respect to the impact of each BOA component, we found that the bidding strategy is by far the most important to consider, followed by the acceptance conditions and finally followed by the opponent model.

The low importance of the opponent model is surprising, as the importance of opponent models has been shown on many occasions. We argue that in our setting, existing learning techniques already do quite well, and consistent with what we found earlier in Chap. 7, no significant effect is to be expected by further improving on currently existing preference learning techniques. This is in contrast to the bidding strategy and acceptance conditions, which have a substantial influence on the agent's performance. To put it another way: our results indicate that the majority of the implementation effort of an agent designer should be focused on the bidding and accepting strategy.

This brings us to the goal of shaping the research agenda on negotiating agent design. Based on our results we recommend that research into bidding strategy and acceptance conditions should stay on the agenda. The state-of-the-art in opponent preference modeling is already so good, that we recommend to focus the attention on the research into automated learning of the bidding strategy and acceptance conditions of the opponent. We will have more to say on this in our overall conclusions in the next chapter.

This chapter is based on the following publications: [16]

Tim Baarslag, Alexander S.Y. Dirkzwager, Koen V. Hindriks, and Catholijn M. Jonker. The signicance of bidding, accepting and opponent modeling in automated negotiation. In *21st European Conference on Articial Intelligence*, volume 263 of *Frontiers in Articial Intelligence and Applications*, pages 27–32, 2014

References

1. Coe R (2002) It's the effect size, stupid: What effect size is and why it is important. In: British Educational Research Association Conference. Education-line
2. Richardson JTE (2011) Eta squared and partial eta squared as measures of effect size in educational research. Educ Res Rev 6(2):135–147
3. Zeng D, Sycara KP (1998) Bayesian learning in negotiation. Int J Human-Comput Stud 48(1):125–141
4. Baarslag T, Fujita K, Gerding EH, Hindriks KV, Ito T, Jennings NR, Jonker CM, Kraus S, Lin R, Robu V, Williams CR (2013) Evaluating practical negotiating agents: results and analysis of the 2011 international competition. Artif Intell 198:73–103

5. Hao J, Leung H-F (2012) ABiNeS: an adaptive bilateral negotiating strategy over multiple items. In: Proceedings of the The 2012 IEEE/WIC/ACM international joint conferences on web intelligence and intelligent agent technology, WI-IAT '12, vol 2. IEEE Computer Society, Washington, pp 95–102
6. Dirkzwager ASY, Hendrikx MJC (2014) An adaptive negotiation strategy for real-time bilateral negotiations. In: Marsa-Maestre I, Lopez-Carmona MA, Ito T, Zhang M, Bai Q, Fujita K (eds) Novel insights in agent-based complex automated negotiation, Japan, studies in computational intelligence, vol 535. Springer, Japan, pp 163–170
7. Faratin P, Sierra C, Jennings NR (1998) Negotiation decision functions for autonomous agents. Robot Auton Syst 24(3–4):159–182
8. Baarslag T, Hindriks KV, Jonker CM (2013) A tit for tat negotiation strategy for real-time bilateral negotiations. In: Ito T, Zhang M, Robu V, Matsuo T (eds) Complex automated negotiations: theories, models, and software competitions, studies in computational intelligence, vol 435. Springer, pp 229–233
9. Morii S, Ito T (2014) AgentMR: concession strategy based on heuristic for automated negotiating agents. In: Marsa-Maestre I, Lopez-Carmona MA, Ito T, Zhang M, Bai Q, Fujita K (eds) Novel insights in agent-based complex automated negotiation, studies in computational intelligence, vol 535. Springer, Japan, pp 181–186
10. van Krimpen T, Looije D, Hajizadeh S (2013) Hardheaded. In: Ito T, Zhang M, Robu V, Matsuo T (eds) Complex automated negotiations: theories, models, and software competitions, studies in computational intelligence, vol 435. Springer, pp 223–227
11. van Galen Last N (2012) Agent Smith: opponent model estimation in bilateral multi-issue negotiation. In: Ito T, Zhang M, Robu V, Fatima S, Matsuo T (eds) New trends in agent-based complex automated negotiations, studies in computational intelligence. Springer, Berlin, pp 167–174
12. Chen S, Weiss G (2014) OMAC: a discrete wavelet transformation based negotiation agent. In: Marsa-Maestre I, Lopez-Carmona MA, Ito T, Zhang M, Bai Q, Fujita K (eds) Novel insights in agent-based complex automated negotiation, studies in computational intelligence, vol 535. Springer, Japan, pp 187–196
13. Williams CR, Robu V, Gerding EH, Jennings NR (2012) Iamhaggler: a negotiation agent for complex environments. In: Ito T, Zhang M, Robu V, Fatima S, Matsuo T (eds) New trends in agent-based complex automated negotiations, studies in computational intelligence. Springer, Berlin, pp 151–158
14. Dirkzwager ASY, Hendrikx MJC, Ruiter JR (2013) The negotiator: a dynamic strategy for bilateral negotiations with time-based discounts. In: Ito T, Zhang M, Robu V, Matsuo T (eds) Complex automated negotiations: theories, models, and software competitions, studies in computational intelligence, vol 435. Springer, pp 217–221
15. Jenks GF, Department of Geography, University of Kansa (1977) Optimal data classification for choropleth maps. Occasional paper. University of Kansas
16. Baarslag T, Dirkzwager ASY, Hindriks KV, Jonker CM (2014) The significance of bidding, accepting and opponent modeling in automated negotiation. In: 21st European conference on artificial intelligence, vol 263 of Frontiers in Artificial Intelligence and Applications, pp 27–32

Chapter 11
Conclusion

Abstract This chapter gives an overview of the contributions of this thesis and revisits and answers the research questions posed in Chap. 1. We outline what we learned through our contributions, we reflect on the impact and limitations of our work, and we describe where we believe research effort should be directed in the following years.

11.1 Contributions

The aim of this thesis project was to research effective ways for an automated negotiating strategy to bid, to learn, and to accept. We presented a method to design and evaluate a component-based automated negotiator, and we introduced new and improved ways for the agent to effectively bid, learn, and to accept.

The contributions to the research into negotiation strategies as presented in this thesis are the following:

1. **Introducing a component-based negotiation architecture to explore the negotiation strategy space** (Chap. 3).

   We introduced a component-based negotiation architecture that distinguishes the bidding strategy, opponent modeling techniques, and acceptance strategies (BOA) in negotiation agents. The BOA framework can be applied in multiple ways to systematically explore the space of automated negotiation strategies. We showed that existing negotiation strategies can be re-fitted into our architecture, and we demonstrated that we can significantly improve the performance of negotiating agents by recombining their components;

2. **Developing evaluation and benchmarking methods for negotiation agents** (Appendices A and B).

   We organized four annual negotiation competitions (ANAC) that had more than 60 international participants in total. The competitions were successful events that have enriched the research field on practical automated negotiation. ANAC

© Springer International Publishing Switzerland 2016
T. Baarslag, *Exploring the Strategy Space of Negotiating Agents*,
Springer Theses, DOI 10.1007/978-3-319-28243-5_11

provides a platform for objectively evaluating generic negotiation agents and makes a wide variety of benchmark negotiation strategies and scenarios available to the research community. ANAC started out as an exploratory search method for generic negotiating agents, but since then, it has matured into a platform that has helped push forward the state-of-the-art in the development and evaluation of automated negotiators.

As our underpinning platform for both ANAC and the BOA framework, we established a negotiation simulation environment (GENIUS) that has proved itself as a valuable and versatile research and analysis tool for the design of generic automated negotiators and for improving their performance. More than one hundred negotiation scenarios are currently available in GENIUS, and the repository of strategies contains more than 40 automated negotiation strategies, including all ANAC 2010–2013 agents. The BOA architecture and all agent components are also included in GENIUS, together with all performance and accuracy measures used in this thesis. The design and evaluation methods of GENIUS have set an accepted standard for the experimental design of automated negotiation, and GENIUS is now used by over 20 research institutes all over the world;

3. **Formulating optimal acceptance policies** (Chaps. 4 and 5).

We determined *optimal* acceptance policies for particular opponent classes using optimal stopping techniques. We classified and compared existing acceptance strategies, and we showed our techniques perform better than the state-of-the-art, demonstrating that the optimal stopping mechanism is a valuable element of a negotiating agent's strategy that works in theory as well as in practice;

4. **Identifying the most effective and accurate learning methods** (Chaps. 6 and 7).

We found out which opponent modeling techniques perform best, why they perform best, and how to best *predict* the performance of opponent models. To do so, we introduced a procedure to evaluate the performance and accuracy of state-of-the-art opponent modeling techniques in negotiation. We determined the best performing opponent modeling techniques and analyzed how the negotiation setting affects the opponent model's effectiveness. We identified the correspondence between accuracy and performance of opponent models, and we found the best methods to predict good performance. We found three measures that are significantly best: *difference in Pareto frontier surface*, *Pearson correlation*, and *ranking distance*.

5. **Formulating optimal bidding strategies and categorizing concession behavior** (Chaps. 8 and 9).

We computed optimal non-adaptive bidding strategies against particular acceptance strategies by employing sequential decision methods. The results show the remarkable fact that the optimal acceptance strategies we found earlier can also be used to make optimal offers.

Moreover, in order to gain insight into the strategy space of bidding techniques, we presented an empirical method to characterize negotiation strategies by their bidding strategy. We used our method to classify agents into four categories of concession behavior (i.e., *Inverter*, *Conceder*, *Competitor*, or *Matcher*), and we formulated guidelines on how agents should bid in order to be successful. The results indicate that in order to be effective, an agent should have a *competitive* negotiation stance;

6. **Quantifying the importance and interactions of the components of a negotiating agent** (Chap. 10).

We found that combining the best agent components indeed improves an agent's overall performance. This validates the analytical approach of the BOA framework towards optimizing the individual components of a negotiating agent. We established that the bidding strategy is by far the most important component for the overall performance of an agent, followed by the acceptance strategy, and finally the opponent modeling component. We also showed how the relative importance of the components depends on the opponent, the characteristics of the negotiation setting and the interaction effect between the components;

7. **Validating the BOA architecture and demonstrating its success**.

In this thesis, the BOA architecture has shown its value in its different applications and was validated in many different ways:

(a) We won first place in the ANAC 2013 competition with an agent (*The Fawkes*) that used the BOA architecture to combine the most effective components. ANAC 2013 had 19 participating negotiation strategies, which were implemented by international experts (Appendix F).

(b) We showed that the BOA framework can be used to design generic, component based negotiation agents by implementing it in GENIUS. The BOA framework now functions as a general software framework and is available as a development tool for agent designers, including participants of ANAC. The BOA framework is also used in education in at least four different international institutes (Chap. 3 and Appendix A).

(c) We demonstrated that existing agents can be re-fitted (i.e., decoupled) into the BOA architecture while retaining exactly the same functionality and with practically the same performance. The BOA framework contains over 90 components from existing agents, which can all be mixed and matched with each other. This shows that the BOA framework is a feasible tool that it is able to incorporate existing agent designs while providing the agent designer with more evaluation and implementation flexibility (Chap. 3).

(d) The BOA framework is an indispensable tool in almost every chapter of this thesis, allowing us to individually optimize the bidding, opponent modeling, and acceptance components of a negotiation agent (Chaps. 4–9).

(e) We showed that indeed, the BOA components can be successfully optimized independently. That is, the BOA components that were best individually, were found to combine into the best agents (Chap. 10).

(f) The BOA framework allowed us to explore the negotiation space and find out which components of a negotiation agent are most important for the overall performance of an agent (Chap. 10).

11.2 Answers to Our Research Questions

All taken together, our contributions outlined in Sect. 11.1 answer the research questions that we originally started with in Sect. 1.4

In our set of *Research Questions I*, we asked how to design a component-based automated negotiation framework, that:

(I.1) supports new agent designs and provides insight into the effectiveness of negotiation strategies;

(I.2) facilitates evaluating and combining various negotiation strategy components;

(I.3) enables us to decompose existing, state of the art agent designs into distinct components.

Question I.1 is amply addressed by *Contribution 1*: by supporting the design of new agents, GENIUS and the BOA framework have been shown to provide insight into the effectiveness of negotiation strategies, not only throughout this thesis, but also in the ANAC competition.

Through our work on the BOA framework (*Contribution 2*) and by implementing it together with GENIUS, we developed an environment that addresses *Question I.2*. The user interface and detailed logging functionality of the BOA framework (including all performance and accuracy measures that we defined in Sects. 2.4.3 and 2.4.4 and all measures we applied in *Contribution 4*) make it easy to evaluate every negotiation component individually. The BOA framework enables us to create thousands of negotiation strategies by combining all negotiation components of the BOA repository.

We decoupled well-known agents from literature, and additionally, by organizing the ANAC competitions (*Contribution 1*), we were able to supplement our repository with more than 60 state of the art negotiation strategies. We decomposed more than 27 of them, resulting in over 90 negotiation components in total (*Contribution 7c*). This shows how we can decompose existing, state of the art agent designs into the BOA framework, which addresses *Question I.3*.

In our set of *Research Questions II* on analyzing the negotiating strategy components, we posed the following questions:

(II.1) What measures can we use to compare and predict the performance of the individual components?

(II.2) Can we pinpoint classes of opponents against which we can find effective components? Can we formulate optimal solutions for any of the components?

(II.3) How does the performance of the components influence the negotiator's performance as a whole, and which components are most important?

With *Contributions 3–6*, we measured the performance of the individual components by employing our component-based BOA architecture (*Contribution 2*), keeping one component fixed, while we varied the others in a negotiation tournaments similar to ANAC (*Contribution 1*). We followed this approach in particular when we quantified the importance of every component (*Contribution 6*). For the bidding strategies and acceptance strategies, we mainly used the average utility as our performance measure, but for the opponent modeling components, we identified an array of performance measures (and additionally, accuracy measures) and formulated ways to predict their performance *(Contribution 4)*. This answers *Question II.1*.

Question II.2 is addressed in a number of contributions throughout this thesis. For the acceptance strategies, we determined a set of effective acceptance strategies for a heterogeneous set of opponents (i.e., the top 3 of ANAC 2010), and we identified optimal acceptance strategies against both unpredictable and time-dependent agents (both described in *Contribution 3*). For the opponent modeling component, we managed to identify the most effective learning methods in *Contribution 4*. Finally, our work on classifying bidding strategies and computing optimal bidding curves (*Contribution 5*) gives us specific guidelines on what kind of bidding behavior to display given the negotiation behavior of the opponent.

Lastly, *Contribution 6* addresses *Question II.3*: by combining the components of a large set of negotiating agents, we established how each of the components contributes to the performance of a negotiator, how they interact with each other, and which component is most important to consider when designing an agent.

11.3 Outlook and Challenges

We conclude this thesis with a general outlook on the topics we have covered, and we present the research challenges that we consider relevant for future work.

11.3.1 The BOA Architecture

The BOA framework has been instrumental in almost every chapter of this thesis, allowing us to individually optimize the bidding, opponent modeling, and acceptance components of a negotiation agent. We showed that these components can be successfully optimized independently to explore the negotiation space.

The BOA architecture has already been widely applied since it was first released. Since its implementation in 2011, the BOA architecture has been used in the ANAC competitions that followed. In ANAC 2012, the BOA agent *The Negotiator Reloaded* reached the finals and finished overall third and received the reward for best performing agent in non-discounted domains. In ANAC 2013, two agents that used the BOA architecture reached the finals, of which *The Fawkes* agent won the 2013 competition.

The BOA architecture has also found its way into the classroom. At academic institutes such as *Bar-Ilan University*, *Ben-Gurion University of the Negev*, *Maastricht University*, and *Delft University of Technology*, GENIUS and the BOA architecture have been integrated into artificial intelligence courses, where part of the syllabus covers automated negotiation and the creation of negotiation strategies.[1]

There are many ways in which the BOA architecture could be extended. One possible improvement is to add additional components to the architecture, by expanding the opponent modeling techniques to a larger class of learning components, such as *strategy prediction*, or *opponent profiling*. Comparing learning techniques for these types of opponent attributes would require more learning techniques to be developed and a new generic negotiation architecture to compare them. Also, BOA agents are currently equipped with a single component during the entire negotiation session. It would be interesting to run multiple BOA components in parallel, and use recommendation systems (or a *meta-component*) to select the best component at any given time. We discuss the components of the BOA architecture in more detail in the sections below.

11.3.2 Bidding

In this thesis, we classified agents according to their concession behavior, and we formulated guidelines on how agents should bid in order to be successful. Our work in Chaps. 8 and 9 makes a number of contributions, but formulating a general and effective bidding strategy for all circumstances is still an important open question in automated negotiation. The importance of finding such a strategy is confirmed by our results in Chap. 10, which show that more than half of the utility obtained by a negotiating agent is determined by its bidding strategy.

A major challenge in defining a generally effective bidding strategy is that there is no universally agreed upon method to define its effectiveness in isolation. In other words, even when we would know exactly when to accept and when we have perfect knowledge about the opponent's preferences, it is still difficult to decide what offers

---

[1] All educational material for the BOA architecture can be freely downloaded from http://ii.tudelft.nl/genius.

to propose. We do not have an established way to measure what a good strategy is, other than to empirically evaluate what works best against a set of opponents. This is in sharp contrast to the opponent modeling component, for which we can employ the accuracy measures from Chap. 7 to quantify their effectiveness. The urgency of the problem is underlined by the fact that even after four incarnations of ANAC, some of the participants (e.g., *HardHeaded*, winner of ANAC 2011) still use concession strategies (e.g. [1]) that were devised in 1998 [2].

Other important concerns when designing the bidding strategy are *multi-level reasoning* and *exploitability*. In order to select the right offer to make, the agent needs to simultaneously answer a variety of questions that not only concern itself, but also the mental state of the opponent; for example: what does the opponent want? What will the opponent learn from the proposed offer? How will the opponent change its mental state, and how will this affect the bids that will be received in the future? Such issues become even more important in a tournament setting, where it is sometimes advisable to sacrifice some utility to the detriment of the other competitors. This makes the bidding strategy the most complex component of the three we distinguish in this thesis.

To make progress in this area, we require a better understanding of all the factors involved in making bids. In particular, we need to understand the interactions between the agent's preferences, the opponent's preferences, the time remaining in the negotiation, the opponent's strategy, and the overall performance measure to optimize (such as tournament score).

As a first step towards understanding these relationships we propose to study *final* offers. Making a final offer in a negotiation (i.e., announcing to 'take it or leave it') is a well-known real-life negotiation tactic that is primarily used as a means to put pressure on the opponent to accept. It also occurs naturally (and more implicitly) in the negotiations in which there is some kind of deadline. Any offer that is made right before the deadline can be considered a final offer, as there may not be enough time to generate a counter-offer. In that case, the opponent is forced to choose between accepting the offer or breaking off the negotiation. This is similar to the ultimatum game, only with a prior bargaining history known to both players.

We believe final offers could act as a useful device to answer some questions about the bidding strategy that would be too difficult to answer in the more general setting, such as: what is the optimal final offer to make at any given point in time, and what is the expected utility? How does it depend on the beliefs about the opponent? In what circumstances would it appropriate to make a final offer?

If we set the total number of rounds to 1, then some of the results we obtained in Chap. 9 translate over to the setting of final offers. We provide a sample result that shows how we might select a final offer given the discount factor and beliefs about the opponent.

Sample result Let the negotiation setting be the *Split the Pie* game with discounted utilities (with discount factor δ), and with the possibility of making
final offers. Agent A and B have reservation values $\text{rv}_A, \text{rv}_B \in [0, 1]$ respectively. Suppose B believes A's reservation value is distributed uniformly. B
contemplates a final offer x (i.e., it would offer $y = 1 - x$ to A, and take x for
himself). The expected utility for B of making a final offer x is:

$$U(x) = x \cdot P(y \geq \text{rv}_A) + \text{rv}_B \cdot P(y < \text{rv}_A)$$
$$= x(y + \text{rv}_B),$$

which is optimal when

$$x_{\text{opt}} = \frac{\text{rv}_B + 1}{2} \quad \text{and} \quad y_{\text{opt}} = \frac{1 - \text{rv}_B}{2},$$

with value $U(x_{\text{opt}}) = x_{\text{opt}}^2$.

Note how this sample result is consistent with our results of Proposition 9.3 for
$j = 1$ on page 176.
As a corollary, when $\delta < x_{\text{opt}}^2$, this gives an equilibrium strategy: B makes the offer
of x_{opt} (and announces the offer as final), and A accepts when $y_{\text{opt}} > \text{rv}_A$.

11.3.3 Opponent Modeling

When it comes to preference learning techniques, it seems the biggest goals have
been accomplished within the scope of this thesis. Simple learning techniques are
already surprisingly effective, even on large domains. We have good measures to
predict the effectiveness of opponent models, and the best techniques are already
at 90% of their upper limit. We also know how the negotiation setting influences
the performance and accuracy. In terms of performance, there is not much to gain
from better preference learning techniques, given their rather minimal performance
increases. This is confirmed by our results in Chap. 10, which show that opponent
modeling is the least important aspect in the design of an effective agent. Some of
the most successful ANAC agents go as far as to not even explicitly model the utility
space of the opponent. Although *Agent K* (winner of ANAC 2010) and *HardHeaded*
(winner of ANAC 2011) still use learning to determine the appropriate concession
level, they do not explicitly model the opponent's utility space.

A possible explanation for the low importance of opponent modeling is the fact that
the number of offers that can be exchanged between the ANAC agents easily reaches

tens of thousands, allowing for exhaustive exploration of all but the largest utility spaces. Such a large number of offers clearly makes it easier for software agents to explore the utility space, yet one could argue whether this is entirely realistic. Even though software agents are able to compute many more offers than humans, in practice there may be other constraints, such as network delays, that would limit the number of offers bargaining agents could exchange. Thus, introducing stochastic break-offs or a minimum time delay between offers may be an issue to consider in our negotiation model.

Some other potential interesting research lines include the following:

1. We need a clearer understanding of how time influences the learning of the opponent's preferences. As we discussed in Chap. 7, many agents take a rather adhoc approach towards the passing of time, causing them to actually become *less* accurate over time. The main cause of this phenomenon is that the bids presented later on in the negotiation are incorrectly handled. The most often used learning techniques treat every received bid the same way, independent of the time it is received [3–5]. To solve this, we need to understand the relationship between the opponent's bidding strategy and the opponent's preferences and how we can learn them in tandem.

2. Many of the learning techniques presented in this thesis require hundreds of bids of input before converging to a reasonable estimate. New methods are required that can learn from a smaller sample of bids if we wish to apply them in other domains, such as human-computer negotiations, continuous domains, and non-linear preferences.[2] For this, we need a better understanding of how different modeling assumptions (as we explored in Chap. 6) affect the faithfulness of the model, and what it means for the assumptions to be robust with respect to a given set of potential opponents.

3. When learning the opponent's preference profile, a learning technique usually makes assumptions about the structure of the negotiation scenario (e.g., [6–8]). Negotiation strategies can exploit the internal structure of the issues of the negotiation domain in order to improve their proficiency. For example, a learning technique can benefit from the information that a certain issue is *predictable*. Informally, an issue is predictable when the global properties of its evaluation function is known. To illustrate, consider the discrete issue *Amount of funding* from the *Zimbabwe–England* domain (cf. Sect. C.2). Its values are: *no agreement*, *$10 billion*, *$50 billion*, or *$100 billion*. Even without any additional information, we can be confident that the utility of each party is either increasing or decreasing in the amount of funding. A price issue like this is typically predictable, but other issues, such as color, are more difficult learn about. Learning to label issues as either predictable or unpredictable could dramatically improve the efficiency of learning algorithms.

---

[2]Finding new learning techniques for the non-linear case is especially relevant for ANAC 2014; see Sect. 11.3.5.

11.3.4   Accepting

We have come a long way in formulating effective acceptance strategies for negotiating agents. If we have a good idea of how many offers we can still expect (possibly
using an estimation of the number of rounds remaining) and of the range of bids
we can expect at every time step (using a strategy prediction mechanism), we can
make the optimal choice of when to accept. The optimal choice is only as good as
the estimations, but we have rather good estimators at our disposal, and the optimal
stopping rules are robust to estimation errors in practice.

The most important limitation of our optimal stopping model presented in Chap. 5
is that the effect of rejecting the opponent's offer is not included in the model. If the
opponent's offer is not accepted, the opponent's behavior is affected by the counter
offer, thus making the bidding strategy an important aspect to consider, as our results
in Chap. 10 on their interaction effects also makes clear.

Second, our model already incorporates the concept of negotiation costs, but we
assumed them to be zero throughout; it would be interesting to see the effects of costs
on optimal acceptance behavior. Similarly, we need new optimal stopping rules for
negotiation scenarios that have discounted payoffs. Both extensions will incentivize
agents to employ more permissive acceptance conditions.

On the other hand, adding reservation values to the agent's preferences would
make an agent *less* inclined to accept. In many cases, it is irrational to withdraw from
a negotiation (i.e., by sending a message ending the negotiation), as it leaves the agent
with nothing. Combining reservation values with discounted scenarios, as is done in
ANAC 2012, cause both contract utility and outside options to devaluate with the
passing of time. In such a setting, novel acceptance conditions are required that give
more consideration to the negotiation timeline. For example, it can be advantageous
for an agent to end the negotiation prematurely and receive its reservation value,
rather than continuing an exchange of offers while the contract diminishes in value.
This adds an additional dimension to the acceptance dilemma, as prolonging the
negotiation does not necessarily increase the agent's chances of a good outcome and
can induce agents to fall back on their reservation value by *ending* the negotiation
prematurely. This 'outside option' gives rise to a new variety of optimal acceptance
strategies that have to make the optimal choice between continuing, accepting, or
walking away.

11.3.5   The Automated Negotiating Agents Competition

Based on the popularity and the lessons we learned from the competition, we believe
that many of our aims regarding ANAC have been accomplished. Recall that we
set out for this competition in order to steer the research in the area of bilateral
multi-issue negotiation and to enable negotiating agents to be evaluated in realistic
environments with a wide variety of opponents and scenarios. The competition has

achieved just that. Since ANAC is designed in such a way that the opponents, as well as the scenarios in which negotiation occurs are unknown in advance, competition participants are compelled to design generic negotiation agents that perform effectively in a variety of circumstances.

Many teams have participated in the four international competitions so far, and we hope that many more will participate in the future [9]. The four incarnations of ANAC have already yielded more than 60 new strategies and scenarios, which provide a comprehensive and freely available repository against which negotiation agents can be benchmarked. This, in turn, allows the negotiation research community to push forward the state-of-the-art in the development and evaluation of automated negotiators and comparison to other automated negotiators.

Since 2010, we have extended GENIUS with all ANAC resources and with the new functionality described in Appendix B (e.g., negotiation strategies, protocols, scenarios, discount factors, and reservation values), the BOA architecture and agent components from Chap. 3, the acceptance strategies from Chaps. 4 and 5, and the performance and accuracy measures described in Chaps. 6 and 7.

However, as with many competitions, ANAC is continually evolving to address new challenges and issues. Given the lessons learned from running ANAC, we intend to eventually introduce several tracks to the competition to model different aspects of the automated negotiation problem, similar to the tracks of the *Trading Agents Competition* (TAC) [10–14]. We see these parallel competition tracks as naturally supporting the different strands of ongoing research in the automated negotiation community.

For ANAC 2014, we intend to introduce non-linear utility functions into the competition. Non-linear utility functions are generally more complex representations of preferences, with many interdependent issues. In such contexts, finding the ideal contract becomes a difficult, nonlinear optimization problem. Selecting bids corresponding to a target utility is difficult already, let alone constructing a model of the opponent utility [15]. With much recent interest in this area [16–20] we expect the non-linear scenarios and negotiation strategies to be a worthwhile addition to our repository.

Finally, we believe our work on ANAC has influence outside the framework of the competition. For example, the relative success of a meta-learning strategy such as *Gahboninho* (see Sect. D.1) shows how second-level adaptivity can pay off, where first-level adaptivity (such as *Nice Tit for Tat Agent*, see Sect. D.1) does not. Where *Nice Tit for Tat Agent* simply adapts to the opponent's behavior, *Gahboninho* has a meta-learning strategy which first tries to establish the *learning behavior* of the negotiation opponent, and then uses this to exploit its opponent. For example, if the opponent is adapting to the strategy of the agent, then *Gahboninho* will be less flexible. If, on the other hand, the opponent does not seem to adapt, *Gahboninho* will be more flexible. In other words, the strategy tries to establish whether its opponent is a teacher or learner, and adapts accordingly (note that this teacher/learner dilemma has been observed in other game-theoretic learning competitions, such as the Lemonade Game [21, 22]). Such an approach has been shown to be successful in the ANAC competition, but also provides useful insights for practical negotiations in general, and is likely to be useful in future research on automated negotiation. In fact, the study of learning and concession behavior of agents in bilateral negotiations provides a natural framework to explore such issues.

11.3.6 Robustness of Negotiation Strategies

The most commonly adopted criterion for evaluating a negotiation strategy is the average utility payoff it can obtain under different negotiation scenarios against other negotiation strategies, as we discussed in Sect. 2.4.3. We used a similar method in many of the experiments in this thesis, ranking the agents using their average performance in a tournament setup. Note that the goal of achieving the highest score in such a tournament is somewhat different to that of reaching the highest score in an individual negotiation.

An alternative criterion to further encourage the development of flexible negotiators would be the total number of games won, instead of the average utility obtained in these games. However, such a criterion would encourage agents to simply beat their opponent, rather than maximize their own utility. This means that agents would be encouraged to get more utility than their opponent at all cost, even if this means reducing their own utility (e.g. by delaying the agreement in the case of a discounted scenario). Such spiteful behavior has also been found in auction bidding [23, 24] and is realistic in some cases, but obviously encouraging such strategies should not be our goal.

Apart from individual performance, a tournament setting also demands that agents take into account their relative performance; i.e., that they are *robust* in the sense of not yielding too much to the other contestants. However, an agent can only control the outcomes of the negotiations it is involved in; it has no control over the negotiations between other agents. Thus, maximizing the utility of each negotiation an agent participates in can be seen as a good approximation that an agent can take in order to maximize its tournament score.

Another parameter that can potentially affect the diversity in outcomes is the specific composition of the opponent pool. We may be interested to know how the winning strategy would change if the tournament size were chosen differently, and especially, if the mix of opponent strategies were different. Moreover, it is natural to ask whether the agents participating in a tournament have an incentive to switch to a different strategy in order to improve their score.

To this end, one can explore the influence of the tournament pool by considering different mixes of opponent strategies. We have performed additional work on this in [25], in which we used the technique of *empirical game theory* as a method for analyzing the results gathered from ANAC. The technique was first developed by [26] to provide insights into the strategies used in the Trading Agent Competition (TAC) and has been shown to be a useful tool in addressing questions about robustness of trading strategies in [27, 28]. Similar techniques have also been used to analyze continuous double auctions [29]. EGT analysis uses the assumption that the strategy

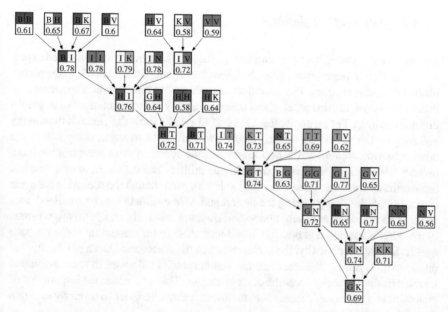

Fig. 11.1 Illustration of deviation analysis for one-to-one negotiations in ANAC 2011. *Arrows* indicate statistically significant reasons for one of the players to switch strategies. At each node, the highest scoring agent is marked by a colored background. Each ANAC 2011 agent is represented by a different letter (*B = BRAMAgent, H = Hardheaded,* and so on)

used by each player is selected from a fixed set of strategies and searches for pure Nash equilibria in a tournament setup.

Specifically, using the payoffs achieved by each agent in a given profile, deviation analysis considers the best single-agent deviation available to an agent in that profile (see Fig. 11.1). A deviation is defined as the incentive of one agent to change its strategy, assuming that all other agents maintain their current strategies. An agent has such an incentive to switch to another strategy if this switch will bring an improvement in its own utility. This approach can be used in order to search for tournaments in which agents have no incentive to deviate (i.e. switch) to a different strategy. Such tournaments are considered to be empirical equilibria. In some games, there may be a subset of profiles, such as the cycle of three shown in Fig. 11.1, each of which are not an equilibrium by themselves, but for which there exists a path of best deviations which connect them, and there is no best deviation which leads to a profile outside of the subset. Such a subset is referred to as a *best reply cycle* [30]. Since an agent never knows which opponent it faces in practice, strategies which are part of an equilibrium, and have a large basin of attraction (i.e., where sequences of deviations often lead to that strategy), are more robust than strategies that are not in equilibrium.

11.3.7 Negotiation Setting

Throughout this thesis, we have emphasized the need for a universally adopted experimental design of negotiation, given how sensitive agent performance is to the particular setup of experiments. The overall scores achieved in one specific tournament, by themselves, do not reveal much about the applicability of the negotiation strategies in different settings. For example, the ANAC 2011 results show that, in both the *Energy* and *Nice or Die* scenarios, agents obtain lower utility on average, compared to the other scenarios. Moreover, the performance of the agents in these scenarios are more diverse, with some agents having much higher utilities than others. In other scenarios, these utilities are much closer. The reason is that, even though the scenarios are quite distinct (with the *Energy* being the largest and *Nice or Die* being the smallest) both of these scenarios have strong opposition (i.e., are relatively competitive), whereas in the other scenarios it is possible to achieve close to the maximum score for both agents. Due to the diversity, the performance in these scenarios has a greater impact on the overall utility, and therefore the strategies which do well in these scenarios have a definite advantage over the other strategies. This underlines the importance of maintaining a variety of scenarios with different characteristics to properly evaluate the performance of an agent.

A possible limitation of negotiating agents is that they often rely on fixed, predetermined parameters in their strategies, such as the time elapsed before the agent becomes more concessive, or the utility an agent should concede to at a given time. Such fixed parameters are used both by less successful agents, such as *ValueModelAgent*, *BRAMAgent* and *The Negotiator*, but also by the ANAC 2011 winner *HardHeaded*. Strategies that try to be more generic, and avoid relying too much on hand-tuned parameters include *IAMhaggler2011*, *Gahboninho* and *Nice Tit for Tat Agent*, but we are still a long way from designing agents that are flexible towards considerable changes in the negotiation setting.

We intend to generalize results of this thesis to settings that deal not only with bilateral, but also with concurrent, one-to-many, and many-to-many negotiations. In this type of negotiation, the concession strategy in each thread may be considerably influenced by the offers made and received in the parallel threads. Such issues have already been explored by the negotiation community, e.g. [31–35], but there is no universally accepted benchmark to compare such agents. Yet another direction of further work could be around *mediated* negotiation scenarios, in which two (or more) negotiating agents incrementally reveal their preferences to a mediator agent that has the task of suggesting a mutually agreeable outcome. Such approaches could be combined with a future track of the ANAC competition that is somewhat similar to the *TAC Market Design competition* and *Power TAC* (see Sect. 2.4.2). Negotiation strategies from previous ANAC competitions could be provided as part of the platform, while participants would be asked to design an agent for this market, or a policy for the mediating agents. A further track could consider a repeated series of negotiations in which the agents can learn based on previous interactions with the opponents.

Another important direction is the design of agents that perform well not only against other automated strategies, but also against human opponents [33, 36–38]. Social interaction, emotions and culture [36] are some of the issues that need to be considered when designing such an agent; issues that were not tackled when designing agents in this thesis. Specifically, different approaches are required, since negotiations with humans need to be much shorter, in terms of the number of offers that can be exchanged in a short timespan.

The results in this thesis are based on discrete domains, where each issue takes a value from a finite set,[3] yet the BOA framework and the GENIUS platform are able to handle continuous issues as well. For future work, we would like to include domains with a combination of continuous and discrete issues. Having continuous issues would generalize the agents even further and benefit application domains where continuous issues occur naturally (such as the allocation of continuous resources, like money or time).

Another important extension is to consider agent utility functions with interdependencies between the issues being negotiated, such as those considered in [39–43]. So far, all the agent utility functions considered in this thesis are additive. However, this may be a limitation since in many real-life scenarios, the utility functions of different agents exhibit complex interdependencies between issues. We have extended the capabilities of GENIUS to incorporate non-linear utility functions for ANAC 2014 (see Sect. 11.3.5).

Another recent development worth noting is the *Negowiki* project [44, 45], which aims to unify current approaches in negotiation research by creating a collection of standardized negotiation scenarios. Their framework is integrated in the *Negowiki* website, where researchers can share and download their scenarios and results. As in GENIUS, analysis of the results is provided, so that researchers can compute a set of metrics over the results of the negotiation. This can work seamlessly in combination with GENIUS: a developer can upload the outcomes of an experiment and the *Negowiki* provides the tools to upload the results to a central repository for others to share. All scenarios offered by *Negowiki* are also available for download in GENIUS format, and non-linear scenarios from *Negowiki* are used for ANAC 2014.

11.3.8 Application to Human Negotiations

An important application of our research would be to introduce it into domains in which one or more of the negotiating parties is human. It would be interesting to evaluate the performance of the various negotiating agents presented in this thesis when they play against, or when they support, human negotiators [46–48]. There are recent extensions of GENIUS that already enable efficient negotiations with human negotiators, using a chat-based interface [49]. The analytical toolbox of GENIUS can

---

[3]With the exception of Chaps. 5 and 9, which formulate optimal bidding and accepting rules for both discrete and continuous cases.

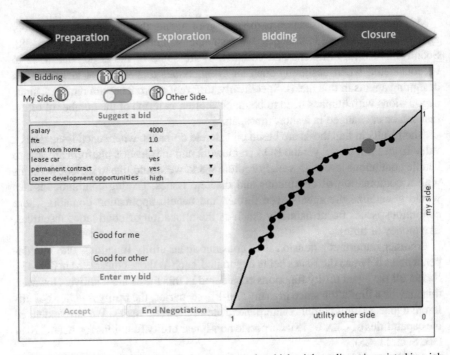

Fig. 11.2 The bidding phase of the *Pocket Negotiator* in which a job applicant is assisted in a job negotiation. Depicted is a suggested bid by the *Pocket Negotiator* (shown in *red*), which is a good offer for the job applicant, but not for the prospective employer. The *black* outcomes constitute the Pareto frontier, and the *green* area in the outcome space indicates the locations of win-win outcomes

be used to discover patterns of negotiation behavior to compare the automated negotiation strategies with human negotiators. Another option is to use GENIUS as a training environment to teach people negotiation concepts, such as exploration of outcome spaces, analysis of opponent's offers, trade-offs between issues, etc.

A first step in applying our research to a real-life setting is the *Pocket Negotiator* [50], of which we have recently released a prototype (Fig. 11.2). The *Pocket Negotiator* is a negotiation support system (NSS) that assists the user in the negotiation process. It can be used to support the user in negotiation with other people and to train human negotiators by means of negotiations against automated agents.

An NSS can help overcome many of the difficulties involved in human negotiations. Humans are usually better at understanding the negotiation context and emotional fluctuations, and have the necessary background knowledge to interpret the negotiation domain. Computers, on the other hand, are capable of storing extensive domain specific knowledge, they can exhaustively search through the entire bid space, and they are not troubled by emotions. Therefore, automated negotiators can work side-by-side humans, where each can benefit from their unique strengths.

It is not easy to create an effective NSS, as the system should be able to be deployed in any negotiation situation, against a beforehand unknown other party. Research into the *Pocket Negotiator* involves many different subprojects and its research challenges include: creating a shared preference model of negotiation [51], considering human values in the design process [52], providing preference elicitation and explanation [53, 54], providing runtime bidding advice [55], and developing affective computing technologies and increasing user awareness of emotions and conflict handling styles [56].

The *Pocket Negotiator* distinguishes four phases of negotiation (seen at the top of Fig. 11.2): preparation, joint exploration, bidding, and closure. The results of this thesis are predominantly used for negotiation support during the bidding phase. To assist the user, the *Pocket Negotiator* can be provided with any of the versatile, automated strategies that were presented in this thesis. When the user requests the *Pocket Negotiator* to suggest a bid, GENIUS is activated, and a negotiation strategy of the designer's choice is consulted for advice.

Our work on the BOA framework is a natural next step in improving the bidding support of the *Pocket Negotiator*. We envision an interface where users can choose the appropriate bidding, learning, and accepting technique to assist them, based on their knowledge of the negotiation domain and the other party.

References

1. Faratin P, Sierra C, Jennings NR (1998) Negotiation decision functions for autonomous agents. Robot Auton Syst 24(34):159–182
2. Baarslag T, Gerding EH, Aydoğan R, schraefel MC (2015) Optimal negotiation decision functions in time-sensitive domains. In: 2015 IEEE/WIC/ACM International joint conferences on web intelligence (WI) and intelligent agent technologies (IAT)
3. Hao J, Leung H (2012) ABiNeS: An adaptive bilateral negotiating strategy over multiple items. Proceedings of the the 2012 IEEE/WIC/ACM international joint conferences on web intelligence and intelligent agent technology, vol 2. IEEE Computer Society, Washington, DC, USA, pp 95–102
4. Hao J, Lung H (2014) CUHK agent: An adaptive negotiation strategy for bilateral negotiations over multiple items. In: Marsa-Maestre I, Lopez-Carmona MA, Ito T, Zhang M, Bai Q, Fujita K (eds) Novel insights in agent-based complex automated negotiation. Studies in computational intelligence, vol 535. Springer, Japan, pp 171–179
5. Hindriks KV, Tykhonov D (2008) Opponent modelling in automated multi-issue negotiation using bayesian learning. Proceedings of the 7th international joint conference on autonomous agents and multiagent systems, International foundation for autonomous agents and multiagent systems, AAMAS '08, vol 1. Richland, SC, pp 331–338
6. Coehoorn RM, Jennings NR (2004) Learning an opponent's preferences to make effective multi-issue negotiation trade-offs. Proceedings of the 6th international conference on Electronic commerce, ICEC '04. ACM, New York, NY, USA, pp 59–68
7. Faratin P, Sierra C, Jennings NR (2002) Using similarity criteria to make issue trade-offs in automated negotiations. Artif Intell 142(2):205–237
8. Zeng D, Sycara KP (1998) Bayesian learning in negotiation. Int J Hum Comput Stud 48(1):125–141
9. Baarslag T, Aydoğan R, Hindriks KV, Fujita K, Ito T, Jonker CM (2015) The automated negotiating agents competition (ANAC). AI Mag 36:2010–2015

10. Greenwald A, Stone P (2001) Autonomous bidding agents in the trading agent competition. IEEE Internet Comput 5(2):52–60
11. Ketter W, Collins J, Reddy P, Flath C, de Weerdt M (2011) The power trading agent competition. ERIM report series reference no. ERS-2011-027-LIS
12. Niu J, Cai K, Parsons S, McBurney P, Gerding EH (2010) What the 2007 tac market design game tells us about effective auction mechanisms. Auton Agent Multi-Agent Syst 21:172–203
13. Stone P, Greenwald A (2005) The first international trading agent competition: Autonomous bidding agents. Electron Commer Res 5(2):229–265
14. Wellman MP, Wurman PR, O'Malley K, Bangera R, de Lin S, Reeves D, Walsh WE (2001) Designing the market game for a trading agent competition. IEEE Internet Comput 5(2):43–51
15. Klein M, Faratin P, Sayama H, Bar-Yam Y (2003) Negotiating complex contracts. Group Decis Negot 12:111–125
16. Ito T, Klein M, Hattori H (2008) A multi-issue negotiation protocol among agents with nonlinear utility functions. Multiagent Grid Syst 4(1):67–83
17. Lopez-Carmona MA, Marsa-Maestre I, Klein M, Ito T (2012) Addressing stability issues in mediated complex contract negotiations for constraint-based, non-monotonic utility spaces. Auton Agent Multi-Agent Syst 24(3):485–535
18. Sánchez-Anguix V, Valero S, Julián V, Botti V, García-Fornes A (2013) Evolutionary-aided negotiation model for bilateral bargaining in ambient intelligence domains with complex utility functions. Inf Sci 222:25–46
19. Sierra C (2012) Negotiation and search. In: AT, p 1
20. Zheng R, Chakraborty N, Dai T, Sycara KP (2013) Multiagent negotiation on multiple issues with incomplete information. In: Proceedings of the 2013 international conference on Autonomous agents and multi-agent systems. International foundation for autonomous agents and multiagent systems, pp 1279–1280
21. Mazliah Y, Gal Y (2005) Coordination in multi-player human-computer groups. In: Proceedings of the first human-agent interaction design and models workshop (HAIDM)
22. Sykulski AM, Chapman AC, de Cote EM, Jennings NR (2010) Ea squared: The winning strategy for the inaugural lemonade stand game tournament. In: Proceedings of the nineteenth european conference on artificial intelligence. Lisbon, Portugal, pp 209–214
23. Brandt F, Sandholm T, Shoham Y (2007) Spiteful bidding in sealed-bid auctions. In: Proceedings of twentieth international joint conference on artificial intelligence, pp 1207–1214
24. Vetsikas IA, Jennings NR (2007) Outperforming the competition in multi-unit sealed bid auctions. In: Proceedings of the 6th international joint conference on autonomous agents and multiagent systems, AAMAS '07. ACM, New York, NY, USA, pp 103:1–103:8
25. Baarslag T, Fujita K, Gerding EH, Hindriks KV, Ito T, Jennings NR, Jonker CM, Kraus S, Lin R, Robu V, Williams CR (2013) Evaluating practical negotiating agents: Results and analysis of the 2011 international competition. Artif Intell 198:73–103
26. Wellman MP, Joshua E, Satinder S, Yevgeniy V, Christopher K, Vishal S (2005) Strategic interactions in a supply chain game. Comput Intell 21(1):1–26
27. Williams CR (Dec 2012) Practical strategies for agent-based negotiation in complex environments. Ph.D. thesis, University of Southampton
28. Chen S, Ammar HB, Tuyls K, Weiss G (2013) Optimizing complex automated negotiation using sparse pseudo-input gaussian processes. Proceedings of the 2013 international conference on autonomous agents and multi-agent systems, AAMAS '13. International foundation for autonomous agents and multiagent systems, Richland, SC, pp 707–714
29. Vytelingum P, Cliff D, Jennings NR (2008) Strategic bidding in continuous double auctions. Artif Intell 172(14):1700–1729
30. Young HP (1993) The evolution of conventions. Econometrica: J Econometric Soc 61(1):57–84
31. Kolomvatsos K, Hadjieftymiades S (2014) On the use of particle swarm optimization and kernel density estimator in concurrent negotiations. Inf Sci 262:99–116
32. Nguyen TD, Jennings NR (2004) Coordinating multiple concurrent negotiations. Proceedings of the third international joint conference on autonomous agents and multiagent systems, AAMAS '04, vol 3. IEEE Computer Society, Washington, DC, USA, pp 1064–1071

33. Traum D, Marsella SC, Gratch J, Lee J, Hartholt A (2008) Multi-party, multi-issue, multi-strategy negotiation for multi-modal virtual agents. In: Prendinger H, Lester J, Ishizuka M (eds) Intelligent virtual agents. Lecture notes in computer science, vol 5208. Springer, Berlin, pp 117–130

34. Williams CR, Robu V, Gerding EH, Jennings NR (2012) Towards a platform for concurrent negotiations in complex domain. In: Proceedings of the fifth international workshop on agent-based complex automated negotiations (ACAN 2012)

35. Williams CR, Robu V, Gerding EH, Jennings NR (2012) Negotiating concurrently with unknown opponents in complex, real-time domains. In: 20th european conference on artificial intelligence

36. Gal Y, Kraus S, Gelfand M, Khashan H, Salmon E (2011) An adaptive agent for negotiating with people in different cultures. ACM Trans Intell Syst Technol 3(1):8:1–8:24

37. Lin R, Kraus S (2010) Can automated agents proficiently negotiate with humans? Commun ACM 53(1):78–88

38. Lin R, Kraus S (2012) From research to practice: automated negotiations with people. In: Krüger A, Kuflik T (eds) Ubiquitous display environments. Cognitive Technologies, Springer, Berlin, pp 195–212

39. Aydoğan R, Baarslag T, Hindriks KV, Jonker CM, Yolum P (2013) Heuristic-based approaches for CP-nets in negotiation. In: Ito T, Zhang M, Robu V, Matsuo T (eds) Complex automated negotiations: theories, models, and software competitions. Studies in computational intelligence, vol 435. Springer, Berlin, pp 113–123

40. Aydoğan R, Yolum P (2012) Learning opponent's preferences for effective negotiation: an approach based on concept learning. Auton Agent Multi-Agent Syst 24:104–140

41. Fujita K, Ito T, Klein M (2009) Approximately fair and secure protocols for multiple interdependent issues negotiation. Proceedings of the 8th international conference on autonomous agents and multiagent systems, International foundation for autonomous agents and multiagent systems, AAMAS '09, vol 2. Richland, SC, pp 1287–1288

42. Marsa-Maestre I, Lopez-Carmona MA, Velasco JR, Ito T, Klein M, Fujita K (2009) Balancing utility and deal probability for auction-based negotiations in highly nonlinear utility spaces. In: Proceedings of the 21st international joint conference on artifical intelligence, IJCAI'09. Morgan Kaufmann Publishers Inc, pp 214–219

43. Robu V, Somefun K, La Poutré JA (2005) Modeling complex multi-issue negotiations using utility graphs. Proceedings of the fourth international joint conference on autonomous agents and multiagent systems. ACM, New York, USA, pp 280–287

44. Marsa-Maestre I, Klein M, de la Hoz E, Lopez-Carmona MA (2011) Negowiki: A set of community tools for the consistent comparison of negotiation approaches. In: Kinny D, Hsu JY, Governatori G, Ghose AK (eds) Agents in principle, agents in practice. Lecture notes in computer science, vol 7047. Springer-Verlag, Berlin, pp 424–435

45. Marsa-Maestre I, Klein M, Jonker CM, Aydoğan R (2013) Towards a negotiation handbook. Decision Support Systems, From problems to protocols

46. Baarslag T, Gerding EH (2015) Optimal incremental preference elicitation during negotiation. In: Proceedings of the twenty-fourth international joint conference on artificial intelligence. AAAI Press, pp 3–9

47. Baarslag T, Gerding EH (2015). Optimal incremental preference elicitation during negotiation. In: 17th international workshop on agent-mediated electronic commerce and trading agents design and analysis

48. Baarslag T, Liccardi I, Gerding EH, Gomer R, Schraefel MC (2015) Negotiating mobile app permissions. In: Amsterdam privacy conference

49. Zuckerman I, Segal-Halevi E, Kraus S, Rosenfeld A (May 2013) Towards automated negotiation agents that use chat interface. In: The sixth international workshop on agent-based complex automated negotiations (ACAN)

50. Jonker CM (2007) The pocket negotiator, synergy between man and machine. NWO Grant proposal

51. Visser W (2012) Qualitative multi-criteria preference representation and reasoning. Dissertation, Delft University of Technology, Delft, The Netherlands
52. Pommeranz A, Detweiler C, Wiggers P, Jonker CM (2012) Elicitation of situated values: need for tools to help stakeholders and designers to reflect and communicate. Ethics Inf Technol 14(4):285–303
53. Pommeranz A, Broekens J, Visser W, Brinkman WP, Wiggers P, Jonker CM (2009) Multi-angle view on preference elicitation for negotiation support systems. Proceedings of the 1st international working conference on human factors and computational models in negotiation. ACM, New York, USA, pp 19–26
54. Pommeranz A (2012) Designing human-centered systems for reflective decision making. Dissertation, Delft University of Technology
55. Tykhonov D (2010) Designing generic and efficient negotiation strategies. Ph.D. thesis, Delft University of Technology
56. Broekens J, Jonker CM, Meyer JJC (2010) Affective negotiation support systems. J Ambient Intell Smart Environ 2(2):121–144

Appendix A
GENIUS: An Environment to Support the Design of Generic Automated Negotiators

Abstract We present an environment called GENIUS, which is a **G**eneral **E**nvironment for **N**egotiation with **I**ntelligent multi-purpose **U**sage **S**imulation. GENIUS helps facilitate both the *design* and *evaluation* of automated negotiators' strategies. It implements an open architecture that allows easy development and integration of existing negotiating agents and can be used to simulate individual negotiation sessions, as well as tournaments between negotiating agents in various negotiation scenarios. GENIUS also allows the specification of different negotiation domains and preference profiles by means of a graphical user interface. GENIUS is employed throughout this thesis as a common environment and testbed to evaluate various ways for automated negotiating strategies to bid, to learn, and to accept. We show the advantages and underlying benefits of using GENIUS and how it can facilitate experimental design in automated negotiation. In particular, it is used as a tournament platform for the negotiation competition discussed in Appendix B.

A.1 Introduction

There are several difficulties that emerge when designing automated negotiating agents, i.e., automated programs with negotiating capabilities. First, while people can negotiate in different settings and domains, when designing an automated agent a decision should be made whether the agent should be a general purpose negotiator, that is, domain-independent (e.g., [2]) and able to successfully negotiate in many settings or suitable for only one specific domain (e.g., the *Colored Trail* domain [3–5], or the *Diplomacy* game [6–8]). There are obvious advantages of an agent's specificity in a given domain. It allows the agent designers to construct strategies that enable better negotiation compared to strategies for a more general purpose negotiator. However, this is also one of the major weaknesses of these types of agents. With the constant introduction of new domains, e-commerce and other applications that require negotiations, the generality of an automated negotiator becomes important, as automated agents tailored to specific domains cannot be re-used in the new domains and applications.

Another difficulty in designing automated negotiators concerns open environments, such as online markets, patient care-delivery systems, virtual reality and

© Springer International Publishing Switzerland 2016
T. Baarslag, *Exploring the Strategy Space of Negotiating Agents*,
Springer Theses, DOI 10.1007/978-3-319-28243-5

simulation systems used for training (e.g., the Trading Agent Competition [9]). These environments lack a central mechanism for controlling the agents' behavior, where agents may encounter opponents whose behavior is diverse.

We do not focus on the design of an efficient automated negotiator here; we do, however, present an environment to facilitate the *design* and *evaluation* of automated negotiators' strategies. The environment, GENIUS, is a **G**eneral **E**nvironment for **N**egotiation with **I**ntelligent multi-purpose **U**sage **S**imulation. To our knowledge, this is the first environment of its kind that both assists in the *design* of strategies for automated negotiators and also *supports* the evaluation process of the agent. Thus, we believe this environment is very useful for agent designers and can take a central part in the process of designing automated agents. While designing agents can be done in any agent oriented software engineering methodology, GENIUS wraps this in an easy-to-use environment and allows the designers to focus on the development of *strategies* for negotiation in an open environment with multi-attribute utility functions.

GENIUS incorporates several mechanisms that aim to support the design of a general automated negotiator; from the initial design, through the evaluation of the agent, to re-design and improvements, based on its performance. The first mechanism is an analytical toolbox, which provides a variety of tools to analyze the performance of agents, the outcome of the negotiation and its dynamics. The second mechanism is a repository of domains and utility functions. Lastly, it also comprises repositories of automated negotiators. A comprehensive description of the tool is provided in Sect. A.2.

In addition, GENIUS enables the evaluation of different strategies used by automated agents that were designed using the tool. The user interacts with GENIUS via a graphical user interface (GUI) and can keep track of the negotiation results with an extensive logging system. This is an important contribution as it allows researchers to empirically and *objectively* compare their agents with others in different domains and settings and validate their results. This in turn allows to generate better automated negotiators, explore different learning and adaptation strategies and opponent models, and collect state-of-the-art negotiating agents, negotiation domains, and preference profiles, and making them available and accessible for the negotiation research community.

We begin by giving an overview of Genius relating to its design.

A.2   The GENIUS System

GENIUS is a **G**eneral **E**nvironment for **N**egotiation with **I**ntelligent multi-purpose **U**sage **S**imulation. The aim of the environment is to facilitate the design of negotiation strategies. Using GENIUS programmers can focus mainly on the strategy design. This is achieved by GENIUS providing both a flexible and easy-to-use environment for implementing agents and mechanisms that support the strategy design and analysis of the agents. Moreover, the core of GENIUS can be incorporated in a larger negotiation support system that is able to fully support the entire negotiation from beginning to end. Examples include the *Pocket Negotiator* [10] and an animated mediator [11]; we give more details in Sect. 11.3.8 of our conclusions.

The design of GENIUS is consistent with our definitions in Sect. 2.2. GENIUS supports arbitrary protocols, with a focus on *bilateral* negotiation. GENIUS can represent arbitrary negotiation *domains*, by allowing the user to define both the negotiation *issues* and the associated range of *values*. A common agent API enables the user to design generic automated negotiators, whose preferences can be prescribed by any given *preference profile*.

In the following sections, we describe the detailed and technical architecture of GENIUS and how it can be used by researchers.

A.2.1 GENIUS' *Architecture*

GENIUS provides a flexible simulation environment. Its architecture, presented in Fig. A.1, is built from several modules: (a) analysis, (b) repository, (c) logging, and (d) simulation control. The analysis module provides researchers the option to analyze the outcomes using different evaluation metrics. The repository contains three different modules of the negotiation that interact with three analysis modules built into GENIUS:

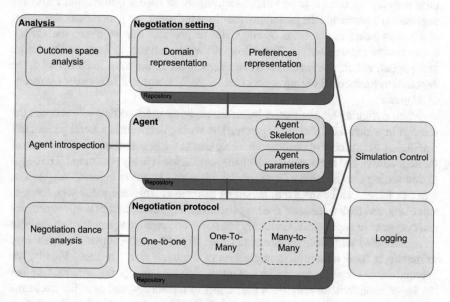

Fig. A.1 The high-level architecture of GENIUS

1. Negotiation scenarios, consisting of a negotiation domain with at least two preference profiles defined on that domain. When a negotiation scenario has been specified, GENIUS is able to perform outcome space analysis on the scenario;
2. Negotiating agents that implement the Agent API. Agent introspection allows the agents to sense the negotiation environment;
3. Negotiation protocols, both one-to-one, and multilateral. Depending on the particular protocol, GENIUS can provide negotiation dance analysis to evaluate negotiation characteristics such as fairness, social welfare, and so on.

Finally, the simulation control and logging modules allow researchers to control the simulations, debug it and obtain detailed information.

A.2.2 GENIUS *as a Tool for Researchers*

GENIUS enables negotiation between automated agents, as well as people. In this section we describe the use of GENIUS prior to the negotiation and afterwards.

Preparation Phase

For automated agents, GENIUS provides skeleton classes to help designers implement their negotiating agents. It provides functionality to access information about the negotiation domain and the preference profile of the agent. An interaction component of GENIUS manages the rules of encounter or protocol that regulates the agent's interaction in the negotiation. This allows the agent designer to focus on the design of the agent, and eliminates the need to implement the communication protocol or the negotiation protocol. Existing agents can be easily integrated in GENIUS by means of adapters.

When designing an automated agent, the designer needs to take into account the settings in which the agent will operate. The setting determines several parameters that dictate the number of negotiators taking part in the negotiation, the time frame of the negotiation, and the issues on which the negotiation is being conducted. The negotiation setting also consists of a set of objectives and issues to be resolved. Various types of issues can be involved, including discrete enumerated value sets, integer-value sets, and real-value sets. The negotiation setting can consist of non-cooperative and cooperative negotiators. Generally speaking, cooperative agents try to maximize their combined joint utilities (see also Chap. 8), while non-cooperative agents try to maximize their own utilities regardless of the other sides' utilities. Finally, the negotiation protocol defines the formal interaction between the negotiators: whether the negotiation is done only once (one-shot) or repeatedly, and how the exchange of offers between the agents is conducted. In addition, the protocol states whether agreements are enforceable or not, and whether the negotiation has a finite or infinite horizon. The negotiation is said to have a finite horizon if the length of every possible history of the negotiation is finite. In this respect, time costs may also be assigned and

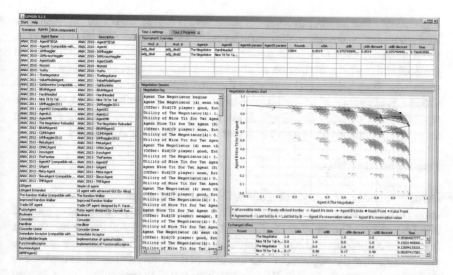

Fig. A.2 An example of GENIUS' main user interface, showing the results of a specific negotiation session

they may increase or decrease the utility of the negotiator. GENIUS provides a testbed which allows the designer to easily vary and change these negotiation parameters.

Using GENIUS a researcher can setup a single negotiation session or a tournament via the GUI simulation (see Fig. A.2) using the negotiation domains and preference profiles from a repository and choose strategies for the negotiating parties. For this purpose, a graphical user interface layer provides options to create a negotiation domain and define agent preferences. This also includes defining different preferences for each role.

A preference profile specifies the preferences regarding possible outcomes of an agent. This can be considered a mapping function that maps the outcomes of a negotiation domain on the level of satisfaction of an agent associated with that outcome. The structure of a preference profile, for obvious reasons, resembles that of a domain specification. The tree-like structure enables specification of relative priorities of parts of the tree. Figure A.3 demonstrates how a preference profile can be modified using GENIUS.

More than one hundred negotiation domains are currently available in the repository of GENIUS. Each domain has at least two preference profiles required for bilateral

Name	Type	Value	Weight
conflict	OBJECTIVE	This == Objective	
Size of Fund	DISCRETE	$100 Billion, $50 Billion, $10 billion, No agreement,	0.3
Impact on Other Aid	DISCRETE	No reduction, Reduction equal to half of fund size, Reduction equal to fund size, No agreement,	0.3
Zimbabwe Trade Policy	DISCRETE	Zimbabwe will reduce tariffs on imports, Zimbabwe will increase tariffs on imports, No agreement,	0.05
England Trade Policy	DISCRETE	England will reduce imports, England will increase imports, No agreement,	0.05
Forum on Other Health Issues	DISCRETE	Creation of fund, Creation of committee to discuss creation of fund, Creation of committee to develop agenda, No,	0.3

Fig. A.3 Setting the preference profile for the *England–Zimbabwe* scenario

negotiations. The number of issues in the domains ranges from 1 to 10, where the largest negotiation domain in the repository is the *AMPO versus City* taken from [12], and has over 7,000,000 possible agreements. Issues in the repository have different predictabilities of the evaluation of alternatives. Issues are considered predictable when even though the actual evaluation function for the issue is unknown, it is possible to guess some of its global properties (for more details, see [13, 14]). The repositories of domains and of agents allow agent designers to test their agents on the different domains and against different kinds of agents and strategies.

Post-negotiation Phase

GENIUS provides an analytical toolbox for evaluating negotiation strategies. This allows to review the performance and benchmark results of negotiators that negotiated using the system. The toolbox calculates optimal solutions, such as the Pareto efficient frontier, Nash product and Kalai-Smorodinsky (see Sect. 2.2.4). These solutions are visually shown to the negotiator or the designer of the automated agent, as depicted in the top right corner of Fig. A.2. We can see all the possible agreements in the domain (all dotted areas), where the highest and most right lines denote the Pareto efficient frontier. During the negotiation each side can see the distance of its own offers from this Pareto frontier as well as the distance from previous offers (as shown by the two lines inside the curve). Also, the designer can inspect both agents' proposals using the toolbox. We note that the visualization of the outcome space together with the Pareto frontier is only possible from the external point of view of GENIUS, which has complete information of both negotiating parties. In particular, the agent themselves are not aware of the opponent utility of bids in the outcome space and do not know the location of the Pareto frontier. The researcher however, is presented the external overview provided by GENIUS that combines the information of both negotiation parties.

Using the analytical toolbox one can analyze the dynamic properties of a negotiation session, with built-in measures such as a classification of negotiation moves (a step-wise analysis of moves) and the sensitivity to a counterpart's preferences measure, as suggested in [13]. For example, one can see whether his/her strategy is concession oriented, i.e., steps are intended to be concessions, but in fact some of these steps might be *unfortunate*, namely, although from the receiver's perception the proposer of the offer is conceding, the offer is actually worse than the previous offer. The result of the analysis can help agent designers improve their agents.

GENIUS keeps track of over 20 different performance measures for the negotiators, such as the *utility performance* of the agents, the *average time of agreement*, and the *percentage of Pareto-efficient bids*. Social welfare measures, such as *average distance from the outcome to the Pareto-frontier, Kalai-point, and Nash-point* are included for all negotiations that result in an agreement. All accuracy measures from Chap. 7 are also implemented in GENIUS, such as *Pearson correlation, ranking distance*, and *average difference* between the real and estimated preferences of an agent.

A.3 Conclusion

This appendix presents a simulation environment that supports the design of generic automated negotiators. The importance and contribution of GENIUS is that it provides, in addition to the design of domain-independent agents, a general infrastructure for defining negotiation scenarios, and for evaluating agents. GENIUS is publicly available[1] and provides researchers a simple and effective tool for designing negotiations' strategies.

Negotiating agents designed using heuristic approaches need extensive evaluation, typically through simulations and empirical analysis, as it is usually hard to predict precisely how the system and the constituent agents will behave in a wide variety of circumstances. To do so, GENIUS provides an environment for the development of a best practice repository for negotiation techniques. Using GENIUS, many new state-of-the-art negotiation strategies have been developed. GENIUS can be used to develop and test agents, and its easy-to-use agent skeleton makes it a suitable platform for negotiating agent development. Moreover, as we show in Appendix B, GENIUS has proved itself as a valuable and extendable research and analysis tool for tournament analysis. GENIUS has the ability to run a wide range of different tournaments, an extensive repository of different agents and domains, and it contains standardized protocols and benchmarks.

This appendix is based on the following publication: [40]
Raz Lin, Sarit Kraus, Tim Baarslag, Dmytro Tykhonov, Koen V. Hindriks, and Catholijn M. Jonker. Genius: An integrated environment for supporting the design of generic automated negotiators. *Computational Intelligence*, 30(1):48–70, 2014

---

[1]http://ii.tudelft.nl/genius.

Appendix B
The Automated Negotiating Agents Competition (ANAC)

Abstract In Appendix A, we described GENIUS, an environment to design and analyze automated negotiators. To compare different negotiation settings, GENIUS requires a variety of different *negotiating agents*, *protocols*, and *scenarios*. With this in mind, we organized ANAC: an international competition based on GENIUS that challenges researchers to develop successful automated negotiation agents for scenarios where there is no information about the strategies and preferences of the opponents. We present an in-depth exposition of the design of ANAC and the key insights gained from four annual International Automated Negotiating Agents Competitions (ANAC 2010–2013). The key objectives of ANAC are to *advance the state-of-the-art* in the area of practical bilateral multi-issue negotiations and to *encourage the design of agents* that are able to operate effectively across a variety of scenarios. We present an overview of the competition, as well as an exposition of general and contrasting approaches towards negotiation strategies that were adopted by the participants of the competition. Based on analysis in post-tournament experiments, we also provide some insights with regard to effective approaches towards the design of negotiation strategies.

B.1 Introduction

From May 2010 to May 2013 we held four instances of the International Automated Negotiating Agents Competition (ANAC)[2] [15, 17–19] in conjunction with the International Conference on Autonomous Agents and Multiagent Systems (AAMAS). This competition follows in the footsteps of a series of successful competitions that aim to advance the state-of-the-art in artificial intelligence (other examples include

---

[2]http://ii.tudelft.nl/anac.

© Springer International Publishing Switzerland 2016
T. Baarslag, *Exploring the Strategy Space of Negotiating Agents*,
Springer Theses, DOI 10.1007/978-3-319-28243-5

the Annual Computer Poker Competition[3] and the various Trading Agent Competitions (TAC) [20]). ANAC focuses specifically on the design of practical negotiation strategies. In particular, the overall aim of the competition is to advance the state-of-the-art in the area of bilateral, multi-issue negotiation, with an emphasis on the development of successful automated negotiators in realistic environments with incomplete information (where negotiators do not know their opponent's strategy, nor their preferences) and continuous time (where the negotiation speed and number of negotiation exchanges depends on the computational requirements of the strategy). More specifically still, the principal goals of the competition include: (i) encouraging the design of agents that can proficiently negotiate in a variety of circumstances, (ii) objectively evaluating different negotiation strategies, (iii) exploring different learning and adaptation strategies and opponent models, and (iv) collecting state-of-the-art negotiating agents and negotiation scenarios, and making them available and accessible as benchmarks for the negotiation research community.

A number of successful negotiation strategies already exist in literature (e.g. [21–25]; see Chap. 2). However, the results of the different implementations are difficult to compare, as various setups are used for experiments in ad hoc negotiation environments [14, 26]. An additional goal of ANAC is to build a community in which work on negotiating agents can be compared by standardized negotiation benchmarks to evaluate the performance of both new and existing agents.

The competition was established to enable negotiating agents to be evaluated in realistic environments and with a wide variety of opponents and scenarios. Moreover, since the opponents, as well as the scenarios in which negotiation occurs are unknown in advance, competition participants are compelled to design *generic* negotiation agents that perform effectively in a variety of circumstances. These agents, together with a wide range of negotiation scenarios, provide a comprehensive repository against which negotiation agents can be benchmarked. This, in turn, allows the community to push forward the state-of-the-art in the development of automated negotiators and their evaluation and comparison to other automated negotiators.

To achieve this, GENIUS was developed, which is the underpinning platform of ANAC that allows easy development and integration of existing negotiating agents. As explained in Appendix A, it can be used to simulate individual negotiation sessions, as well as tournaments between negotiating agents in various negotiation scenarios.

With GENIUS in place, we organized ANAC with the aim of coordinating the research into automated agent design and proficient negotiation strategies for bilateral multi-issue closed negotiation, similar to what the TAC [20] achieved for the trading agent problem. We believe ANAC is an important and useful addition to existing negotiation competitions, which are either aimed at human negotiations or have a different focus, as we explained in Sect. 2.4.2.

In this appendix, we will outline the general rules, goals and results of the ANAC installments. For specific information on agents, scenarios, and scores, we refer to the Appendices C through F.

---

[3]http://www.computerpokercompetition.org.

ANAC is held in conjunction with AAMAS, which is a well-suited platform to host the competition, as it is the premier scientific conference for research on autonomous agents and multiagent systems, which includes researchers on automated negotiation. It brings together an international community of researchers that are well-suited to tackle the automated agents negotiation challenges posed by ANAC.

This appendix is organized as follows. Section B.2 provides an overview over the design choices for ANAC, including the model of negotiation, tournament platform and evaluation criteria. The participating agents, the scenarios, and the results of four years of ANAC are described in Sect. B.3. Section B.4 closes off with a discussion of the implications of ANAC.

B.2   General Design of ANAC

Our aim in designing ANAC is to provide a strategic challenge on multiple accounts. We begin by describing the challenges set forth by ANAC, and the negotiation model that is used during each negotiation encounter. After that, we describe how the competition tournament is formed as a series of such encounters.

One of the goals of ANAC is to encourage the design of agents that can negotiate in a variety of circumstances. This means the agents should be able to negotiate against any type of opponent within arbitrary domains. Such an open environment lacks a central mechanism for controlling the agents' behavior, and the agents may encounter different types of opponents with different characteristics. Therefore, the participating automated negotiation agents should be capable of negotiating proficiently with opponents that are diverse in their behavior and negotiate in a different manner.

The negotiation model behind ANAC is in line with our definitions in Sect. 2.2: we consider *bilateral* negotiations using the alternating-offers protocol, in which the utilities of the players are *additive*. The design of the competition was focused on the development of negotiating strategies, rather than other aspects of the negotiation process (though not less important aspects) such as preference elicitation, argumentation or mediation. The setup of ANAC was designed to make a balance between several concerns, including:

- Strategic challenge: the game should present difficult negotiation domains in a real-world setting with real-time deadlines.
- Multiplicity of issues on different domains, with a priori unknown opponent preferences.
- Realism: realistic domains with varying opponent preferences.
- Clarity of rules, negotiation protocols, and agent implementation details.

B.2.1 Tournament Platform

As a tournament platform to run and analyze the negotiations, we use the GENIUS environment.[4] GENIUS is a research tool for automated multi-issue negotiation, that facilitates the design and evaluation of automated negotiators' strategies. It also provides an easily accessible framework to develop negotiating agents via a public API. This setup makes it straightforward to implement an agent and to focus on the development of strategies that work in a general environment.

Each participating team has to design and build a negotiation agent using the GENIUS framework. GENIUS incorporates several mechanisms that support the design of a general automated negotiator for ANAC. The first mechanism is an analytical toolbox, which provides a variety of tools to analyze the performance of agents, the outcome of the negotiation and its dynamics. The second mechanism is a repository of scenarios. Lastly, it also comprises repositories of automated negotiators. In addition, GENIUS enables the evaluation of different strategies used by automated agents that were designed using the tool. This is an important contribution as it allows researchers to empirically and *objectively* compare their agents with others in different domains and settings.

As we mentioned in Sect. A.2, the GENIUS framework provides skeleton classes to facilitate the design of negotiating agents. Other aspects of negotiation—specifying information about the domain and preferences, sending messages between the negotiators while obeying a specified negotiation protocol, declaring an agreement—is handled by the negotiation environment. This allows the agent's designer to focus on the implementation of the agent. The agent's designer only needs to implement an agent interface provided by the GENIUS framework. In essence, the agent's developer implements two methods: one for receiving a proposal, and one for making a proposal. The rest of the interaction between the agents is controlled by GENIUS.

GENIUS is freely available to the ANAC participants and researchers to develop and test their agent. Table B.1 gives an overview of the most important information that was available to the agent through the API provided by GENIUS.

The flexibility provided by the built-in general repository makes GENIUS an effective tournament platform. The contestants of ANAC are able to upload their agent source code (or even compiled code) to the ANAC organizers. The agents are then added to the GENIUS repository. The ANAC agents and domains are bundled in the GENIUS repository and released to the public after the tournament. GENIUS also provides a uniform, standardized negotiation protocol and scoring system, as every developing team implements the agent inside the same GENIUS environment.

GENIUS supports a number of different protocols, such as the alternating offers protocol, one-to-many auctions, and many-to-many auctions. See Fig. B.1 for an overview of the types of tournaments that can be run.

The analytical toolbox of GENIUS (see Fig. B.2) provides a method to evaluate the negotiation strategies employed by the ANAC participants. The toolbox gives valuable graphical information during the negotiation sessions, including: Pareto

---

[4]See Appendix A and http://ii.tudelft.nl/genius.

Table B.1 Highlighted functionality of the API available to the agent in order to access information about the negotiation environment and its preferences

Agent	IssueDiscrete (implements Issue)
Action chooseAction()	String getDescription()
Enables the agent to offer a bid to the opponent	Returns a short description of the issue
String getName()	String getName()
Returns the name of the agent	Returns the name of the issue
Timeline getTimeline()	List<ValueDiscrete> getValues()
Gets information about possible time constraints	Returns all values associated with this issue
Double getUtility(Bid bid)	**Timeline**
Computes the *discounted* utility of a bid, given the current Timeline	
UtilitySpace getUtilitySpace()	Double getElapsedSeconds()
Gets the preference profile of the agent	Returns the seconds that have elapsed since the start of the negotiation
receiveMessage(Action opponentAction)	Double getTime()
Informs the agent about the opponent's action	Gets the normalized elapsed time in [0, 1]
Bid	**UtilitySpace**
Value getValue(Issue issue)	Double getDiscountFactor()
Returns the selected value of a given issue in the current bid	Gets the discount factor
setValue(Issue issue, Value value)	Double getReservationValue()
Sets the value of an issue	Gets the agent's reservation value
Domain	Double getUtility(Bid bid)
	Computes the utility of a given bid
List<Issue> getIssues()	**ValueDiscrete** (implements Value)
Gets all issues of the negotiation domain	
	String getValue()
	Returns the text representation of this value

optimal solutions, Nash product, Kalai-Smorodinsky. The negotiation logging system gives insight into the agent's reasoning process and can help improve the agent code. When a particular negotiation has finished, an entry is added to the tournament overview, containing information about the number of rounds used, and both utilities associated with the agreement that was reached by the agents. This information can be used to assess the optimality of the agreements reached, either for both agents, or for each agent individually. The result of the analysis can help new agent designers to improve their agents as they play against previous ANAC strategies.

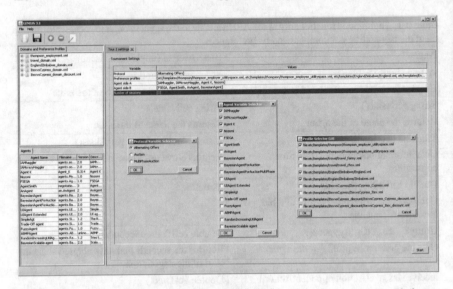

Fig. B.1 Setting up a tournament session for ANAC 2010 involves choosing a protocol, the participating agents, and appropriate preference profiles

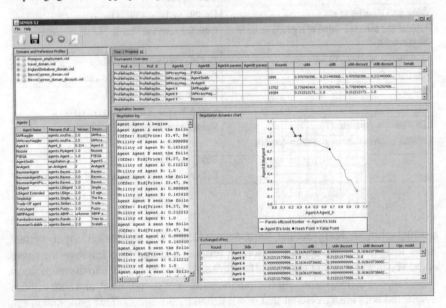

Fig. B.2 A tournament session with the ANAC 2010 agents playing on the ANAC 2010 scenarios using the GENIUS interface

B.2.2   Competition Scenarios

The competition is targeted towards modeling multi-issue negotiations in uncertain, open environments, in which agents do not know the preferences of their opponent. The various characteristics of a negotiation scenario such as size, number of issues, opposition and discount factor can have a significant influence on the negotiation outcome (see also Sects. 2.2.3 and 2.2.4). Due to the sensitivity to the negotiation specifics, negotiation strategies have to be assessed on negotiation domains of various sizes and of various complexity [14].

Therefore, in order to ensure a good spread of negotiation characteristics, and to reduce any possible bias on the part of the organizers, we gathered the domains and profiles from the participants in the competition.[5] Specifically, in addition to submitting their agents, each participant submitted a scenario, consisting of both a domain and a pair of utility functions. We then use the scenarios submitted by the participants to run the tournament.

In all years, the scenarios are unknown to the agents prior to the tournament and the agents receive no information about their opponent's preferences during the tournament. Because ANAC is aimed towards multi-issue negotiations under uncertainty in open environments, we encourage participants to submit domains and profiles with a good spread of the relevant parameters, such as the number of issues, the number of possible proposals and the opposition of the domain. Another degree of uncertainty is the strategies used by the opponents. Thus, although it is possible to learn from the agents and domains of previous years, successful agents submitted to ANAC have to be flexible and domain-independent.

The domains and utility functions used during the competition were not known in advance and were designed by the participants themselves. Therefore, in a given negotiation, an agent does not know the utility function of its opponent, apart from that the fact that it is additive. In more detail, the participants have no prior knowledge of the distribution over the function's parameters and, furthermore, they do not even know the opponent's preference ordering over the values for an individual issue. The pairs of utility functions which form a scenario are designed by the participants (who also develop the agents), but the rules prohibit designing an agent to detect a particular scenario (and therefore the opponent's utility function) based on knowledge of such a pair of functions. Furthermore, the scenarios were changed during the updating period so that the finalists would not benefit from tuning their strategies to the scenarios of the qualifying round.

We approach the overall design of ANAC to comply with the goals that were described in Sect. B.1.

---

[5]For ANAC 2010, there was only one final round, and the domains and preference profiles used during the competition were designed by the organizing team.

B.2.3   Running ANAC

Due to the constant evolving nature of ANAC, the specific ANAC setup and rules are slightly different every year; however, in general, the timeline of ANAC consists of three phases: the qualifying round, the updating period and the final round. An agent's success is measured according to the average utility achieved in all negotiations of the tournament for which it is scheduled.

First, a *qualifying round* is played in order to select the best 8 agents from the agents that are submitted by the participating teams. Since ANAC 2010, each participant also submits a domain and pair of utility functions for that domain. All these scenarios are used in the qualifying rounds. For each of these scenarios, negotiations are carried out between all pairings of the agents. The 8 agents that achieve the best average scores during qualifying are selected as participants for the final round.

Our evaluation metric is defined as follows. Every agent A plays against all other agents, but A will not play itself. The score for A is averaged over all trials, alternating between the two preference profiles defined on every domain (see Sect. B.2.2). For example, on the *Itex versus Cypress* domain (Table C.3), A will play both as *Itex* and as *Cypress* against all others. Note that these averages are taken over all negotiations, excluding those in which both agents use the same strategy (i.e. excluding self-play). Therefore, the average score $U_\Omega(A)$ of agent A in scenario Ω is given formally by:

$$U_\Omega(A) = \frac{\sum_{B \in P, A \neq B} U_\Omega(A, B)}{(|P| - 1)} \tag{B.1}$$

where P is the set of players and $U_\Omega(A, B)$ is the utility achieved by player A against player B in scenario Ω.

Note that with the exception of ANAC 2013,[6] we only consider situations where the player can perform this learning *within* a negotiation session and that any learning cannot be used *between* different negotiation encounters. This is done so that agents need to be designed to deal with unknown opponents. In order to prevent the agents learning across instances, the competition is set up so that a new agent instance is created for each negotiation. The rules prohibit the agents storing data on disk, and they are prevented from communicating via the Internet.

Agents can be disqualified for violating the spirit of fair play. The competition rules allowed multiple entries from a single institution but required each agent to be developed independently. Furthermore it was prohibited to design an agent that benefits some other specific agent (c.f. the work on collusion in the Iterated Prisoner's Dilemma competitions in 2004 and 2005 [27]).

---

[6]For ANAC 2013, we allowed partial learning across domains, as described in Sect. B.3.4.

B.3   The ANAC Installments

Every year, new features are incorporated into the competition environment to increase realism and to encourage the development of flexible and practical negotiation agents. After every ANAC, the participating teams have a closing discussion, yielding valuable suggestions for improving the design of following ANAC competitions, introducing small innovations every year. We shortly describe four years of ANAC, with a focus on ANAC 2010 and 2011, since their results and resources are used most often in this thesis. For more details, we refer to Appendices C–F.

B.3.1   ANAC 2010

ANAC started in 2010 as a joint project of the universities of Delft and Bar-Ilan and had seven participating teams from five different universities, as listed in Table C.1 of Appendix C. It was held at the Ninth International Conference on Autonomous Agents and Multiagent Systems (AAMAS 2010) in Toronto, Canada, with presentations of the participating teams and a closing discussion.

ANAC 2010 was the only instance of ANAC without a qualifying round and in which the organizers selected the negotiation scenarios instead of the participants. Three different scenarios were selected, which can be viewed in Table C.3 and are visually depicted in Fig. C.1.

The scores of every agent in ANAC 2010 are listed in Table C.2. *Agent K* won ANAC 2010 by a relatively large margin, yet it only managed to dominate on the *Travel* domain. On both *Itex versus Cypress* and *England–Zimbabwe* scenarios, it earned second place after *Nozomi* and *Yushu*, respectively. However, *Agent K* won the competition due to its consistent high scores in all domains. Only *IAMhaggler* managed to mirror this consistent scoring on all three domains.

Note that for ANAC 2010 only, the final scores are normalized per domain. The utility is normalized for every profile, using the maximum and minimum utility achieved by all other agents. This gives a score per profile, which is averaged over the two profiles in the domain to give an overall normalized domain score. The domain score is then averaged over all trials and yields the final score of the agents. Due to the normalization of the scores, the lowest possible score is 0 and the highest is 1 for every domain. The fact that the maximum and minimum score are not always achieved by ANAC 2010 agents is due to the non-deterministic behavior of the agents: the top ranking agent on one domain does not always obtain the maximum score on every trial.

Table B.2 gives a more detailed overview of the strategy of all agents. Note that most of the agents are non-deterministic, which is relevant for the experiments in this thesis since it introduces noise in the tournament results. To illustrate: during a negotiation, *Agent K* may decide on a certain proposal target. But if it previously received better offers *B*, then it will counteroffer a random offer taken from *B*.

Table B.2 Strategy details of the agents that participated in ANAC 2010

Agent	Time dependent	Learning method	Acceptance criteria	Deterministic
Agent K	Yes	All proposals	Time/utility	No
Yushu	Yes	Best proposals	Time/utility	No
Nozomi	No	Match compromises	Time/utility	No
IAMhaggler	Yes	Bayesian learning	Utility	No
FSEGA	Yes	Bayesian learning	Utility	Yes
IAMcrazyHaggler	No	None	Utility	No
Agent Smith	Yes	Learn weights	Time/utility	Yes

Listed details are: (1) whether the strategies change their proposals according to the remaining time; (2) what kind of learning method is used; (3) whether the agents take the offer's utility or remaining time into account when accepting; (4) whether the agents are deterministic

Otherwise, it will also select a random proposal; in this case it will choose any offer that satisfies its proposal target. Most agents have a similar mechanism, which we elaborate on in Chap. 3: when they are indifferent between certain offers, they will choose randomly.

All agents of ANAC 2010, except for *IAMcrazyHaggler*, make concessions when the deadline approaches. Because a break-off yields zero utility for both agents, an agent that waits until the end of the negotiation takes a substantial risk. The other agent may not know that the deadline is approaching and may not concede fast enough. In addition, either the acceptance of a proposal or the (acceptable) counter-offer may be received when the game is already over. In the same manner, a real-time deadline also makes it necessary to employ a mechanism for deciding when to accept an offer.

We study the inclination of the agents of ANAC to exhibit either risk averse or risk seeking behavior in more detail in Chap. 8. In order to get a good picture of the risk management of the agents, we consider here the number of break-offs that occur for every agent. Table B.3 lists for each agent the percentage of negotiations

Table B.3 Percentage of all failed negotiations of every agent per domain

Agent	Break-off percentage			
	Itex–Cyp (%)	Eng–Zimb (%)	Travel (%)	Avg. (%)
Agent K	22	6	63	30
Yushu	36	0	90	42
Nozomi	25	17	75	39
IAMhaggler	11	0	63	25
FSEGA	22	0	100	41
IAMcrazyHaggler	72	23	83	59
Agent Smith	0	0	98	33

that result in a break-off. All break-offs occur due to the deadline being reached or an occasional agent crash on a big domain.

The number of break-offs in the *Travel* domain stands out compared to the other domains. Recall that this is the biggest domain of ANAC 2010, with 188,160 possible proposals. Most of the agents had problems dealing with this domain. With such a large domain, it becomes unfeasible to enumerate all proposals or to work with an elaborate opponent model. For example the *FSEGA* agent was unable to finish a single negotiation. Only *Agent K*, *Nozomi* and *IAM(crazy)Haggler* were able to effectively negotiate with each other on this domain, which resulted in less break-offs for them, hence their higher scores.

With respect to the number of break-offs, *IAMHaggler* performs very well on all domains, while *IAMcrazyHaggler* ranks as the worst of all agents. This is to be expected, as its proposal generating mechanism does not take into account the time or the opponent (see Sect. C.1 for an overview of its strategy). There is an interesting trade-off here: when *IAMcrazyHaggler* manages to reach an agreement, it always scores a utility of at least 0.9, but most of the time it scores 0 because the opponent will not budge.

The exact opposite of *IAMcrazyHaggler* is the strategy of *Agent Smith*. Because of an implementation error, *Agent Smith* accepts any proposal after two minutes, instead of three minutes. This explains why it did not have any break-offs on *Itex versus Cypress* and *England–Zimbabwe*. The reason for the break-offs on the *Travel* domain is due to crashing of its opponent model. The importance of the timing aspects is underlined by the performance of *Agent Smith*: a small timing error resulted in very poor scoring on all three domains.

B.3.2   ANAC 2011

The local organization of ANAC 2011 was in the hands of the winner of 2010, Nagoya Institute of Technology. Eighteen teams (as compared to seven in the first competition) submitted negotiating agents to the tournament, which was held during the Tenth International Conference on Autonomous Agents and Multiagent Systems (AAMAS 2011) in Taipei, Taiwan. The teams came from seven different institutes (University of Alcalá, Bar-Ilan University, Ben-Gurion University, Politehnica University of Bucharest, Delft University of Technology, Nagoya Institute of Technology, and University of Southampton) and six different countries (Spain, Israel, Romania, the Netherlands, Japan, and the United Kingdom).

In contrast to the first competition, ANAC 2011 introduced a discount factor for some of the scenarios, to incentivize the agents to have more interesting negotiations with faster deals (see Sect. 2.2.3). Agents still needed to operate in both discounted and undiscounted settings: the discount factor was disabled (i.e., equal to 1) for half of the scenarios; for the other half, the discount factor was decided randomly.

In ANAC 2010, the agents had three minutes *each* to deliberate. This meant the agents had to keep track of both their own time and the time the opponent had left.

For ANAC 2011 and onwards, we elected a simpler protocol where both agents have a shared time-line of three minutes. This means that, if one agent causes a delay, this will affect both agents equally, both in terms of the discounting and getting closer to the deadline.

This time, the participants submitted one agent and one negotiation scenario. Eight of these teams continued to the finals after undergoing a qualifying round (see Table D.1 of Appendix D). The qualifying round consisted of the 18 agents that were submitted to the competition. For each pair of agents, under each utility function, we ran a total of 3 negotiations. By averaging over all the scores achieved by each agent (against all opponents and using all utility functions), eight finalists were selected based on their average scores.

Between the rounds, we allow a number of weeks as an updating period, in which the 8 selected finalists were given the chance to improve their agents for the final round. The detailed results and all scenarios for the qualifying round were revealed to all finalists, and they could use this additional information to tune their agents.

Since there were 18 agents, which each negotiate against 17 other agents, in 18 different domains, a single tournament in the qualifying round consists of $18 \times 17/2 \times 2 \times 18 = 5508$ negotiation sessions.[7] To reduce the effect of variation in the results, the tournament was repeated 3 times, leading to a total of 16,524 negotiation sessions, each with a time limit of three minutes. In order to complete such an extensive set of tournaments within a limited time frame, we used five high-spec computers, made available by Nagoya Institute of Technology. Specifically, each of these machines contained an *Intel Core i7* CPU and at least 4GB of DDR3 memory. Allocating the entire tournament took one month to run.

It is notable that *Gahboninho* was the clear winner of the qualifying round (see Table D.1). As we discuss in Chap. 8, we believe its strong performance is partly due to the learning approach it adopts, in an attempt to determine whether the opponent is cooperative.

The tournament among 8 finalists was played on the 8 scenarios submitted by all finalists (cf. Table D.3). The entire set of pairwise matches were played among 8 agents, and the final ranking of ANAC 2011 was decided. We matched each pair of finalists, under each utility function, a total of 30 times. In the final, a single tournament consists of $8 \times 7/2 \times 2 \times 8 = 448$ negotiation sessions. Table D.2 summarizes the means, standard deviations, and 95 % confidence interval bounds for the results of each agent, taken over the 30 iterations.[8] In common with the approach used in the qualifying round, all agents use both of the profiles that are linked to a scenario. Note the small differences of the scores of the agents in positions 4 to 7. Specifically,

---

[7]The combinations of 18 agents are $18 \times 17/2$, however, agents play each domain against each other twice (once for each profile).

[8]The standard deviations and confidence intervals are calculated based on the variance of the utilities across the 30 iterations of the tournament (after being averaged over all of the scenarios). Therefore they only measure the variance across complete tournaments, which may be due to intentional randomness within the agents' strategies or stochastic effects that are present in the tournament setup.

there is no statistically significant difference between the utilities achieved by *Agent K2, The Negotiator, BRAMAgent,* and the *Nice Tit for Tat Agent.*

The shape of the outcome space of each scenario is presented graphically in Figs. D.1 and D.2. In more detail, very large scenarios, such as the *Energy* scenario, are displayed with a large number of points representing the many possible agreements, whereas smaller scenarios, such as *Nice Or Die,* have only very few points. Furthermore, scenarios which have a high mean distance to the Pareto frontier, such as the *Grocery* and *Camera* scenarios, appear very scattered, whereas those with a low mean distance, such as the *Company Acquisition* scenario, are much more tightly clustered. The other 10 scenarios (which were eliminated along with their agents in the qualifying round) contained broadly similar characteristics to those of the final 8 scenarios. Therefore, since the final 8 scenarios capture a good distribution of the characteristics we would like to examine, we consider only these scenarios in the rest of this thesis.

The average score achieved by each agent in each scenario is given in Table B.4, and presented visually in Fig. B.3. In the finals, *HardHeaded* proved to be the winner,[9] with a score of 0.749. Figure B.3 clearly shows that in most scenarios the margin between the worst and the best agents was minimal. Specifically, in 6 of the 8 scenarios, the worst agent achieved no less than 80% of the best agent's score. The remaining two scenarios that had a much greater range of results were also the scenarios with the greatest opposition between the two utility functions. We see that winning the competition does not require the agent to win in all or even most scenarios. The results presented in Table B.4 and Fig. B.3 show that the winning *Hard-Headed* agent did not win in the majority of the scenarios (it only did so in 3 out of 8). The runner-up, *Gahboninho,* had the highest utility in 2 of the scenarios in the finals. As long as an agent wins by a large margin in those scenarios where it comes first, it can win the entire competition. *IAMhaggler2011* won the *Company Acquisition* and *Laptop* scenarios where there is a low discount factor; therefore, *IAMhaggler2011* is well suited to cases where agreements need to be reached quickly. Its high degree of adaptivity enables the agent to reach efficient agreements, even in large domains, or in scenarios that are subject to considerable time discounting. However, while *IAMhaggler2011* performed well in general, this did not secure it a winning position overall, nor in many of the specific scenarios. The *Nice Tit for Tat Agent* performed rather poorly overall. The failure of this agent is due to its relative cooperativeness, i.e. willingness to adapt to the opponent's demands. We come back to this kind of reasoning in Chap. 8.

Interestingly, the results show no clear connection between the discount factor and the diversity in performance. This could be due to the fact that the most discounted scenarios (i.e. *Laptop* and *Company Acquisition,* which have the lowest discount

---

[9]There are a number of reasons why the winner in the final round was different to the qualifying round. Firstly, the set of scenarios used in the final was smaller than in the qualifying round, and it is possible that the final scenarios were more favorable to the *HardHeaded* agent. Secondly, the set of participating agents was smaller, and furthermore, due to the elimination of the lower scoring agents, those agents that remain were more competitive. Finally, it is possible that the agents were modified between the two rounds.

Table B.4 Detailed scores of every agent in each scenario in the final round of ANAC 2011

Agent	Nice or Die	Laptop	Company acquisition	Grocery	Amsterdam	Camera	Car	Energy
HardHeaded	**0.571**	0.669	0.749	0.724	0.870	**0.811**	**0.958**	0.637
Gahboninho	0.546	0.730	0.752	0.668	**0.929**	0.665	0.946	**0.682**
IAMhaggler2011	0.300	**0.750**	**0.813**	0.726	0.781	0.715	0.864	0.543
Agent K2	0.429	0.655	0.788	0.717	0.750	0.727	0.921	0.459
The Negotiator	0.320	0.651	0.757	0.733	0.791	0.742	0.930	0.519
BRAMAgent	**0.571**	0.631	0.747	0.725	0.792	0.739	0.803	0.432
Nice Tit for Tat Agent	0.425	0.668	0.772	0.753	0.739	0.774	0.786	0.509
ValueModelAgent	0.137	0.641	0.764	**0.765**	0.857	0.781	0.951	0.037

Bold text is used to emphasize the best score achieved in each scenario

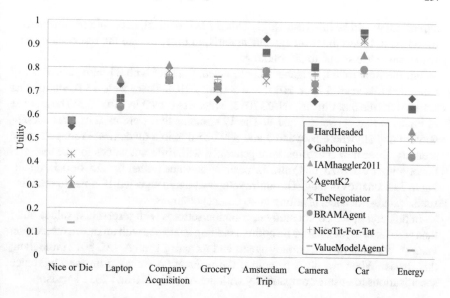

Fig. B.3 Scores of every agent in each scenario in the final round of ANAC 2011

factor) also have weak opposition and small domains, meaning that win-win agreements can be easily found. It would be interesting to apply the discount factor to other types of scenarios as well (i.e., with larger domains and stronger opposition), to see the impact of the discount factor in more challenging settings.

B.3.3 ANAC 2012

For ANAC 2012, 17 teams entered from 8 different institutions from 5 countries (China, Israel, Netherlands, Japan, United Kingdom); see Table E.1. The University of Southampton was the local organizer of this tournament, which was co-located with the Eleventh International Conference on Autonomous Agents and Multiagent Systems (AAMAS 2012) in Valencia, Spain.

For the qualifying round, negotiations were carried out for every combination of the 17 participants on 18 negotiation scenarios (17 submitted this year, plus the *Travel* domain from 2010). Each negotiation was repeated 10 times to establish statistical significance, which resulted in a total of 52020 negotiations.

In 2012, the competition introduced a private reservation value as part of the tournament (see Sect. 2.2.3). When an agent failed to reach an agreement by the deadline, or if one of the agents terminated the negotiation, both received their reservation value instead of zero utility. The reservation value could be different for each agent and for each negotiation scenario. Each agent only knew its own reservation value, and not that of its opponent. The reservation value was discounted in the same way that an agreement would be in ANAC 2011. This made it rational, in certain circumstances, for an agent to terminate an agreement early, in order to take the reservation value with a smaller loss due to discounting.

For the final round, 8 agents were selected, together with 24 base scenarios (17 submitted this year, 5 from 2011, and 2 from 2010). There were 13 scenarios that featured for the first time in ANAC 2012, which are shown in Table E.3. The number of issues varied between 1 and 8. The 17 domains that were used to run the competition had anywhere between 3 and 390,625 possible outcomes. From each base scenario, three new scenarios were generated with different values for the discount factor (either 0.5, 0.75, or 1) and for reservation value (either 0, 0.25, or 0.5), resulting in 72 scenarios in total. The entire setup was again repeated 10 times to establish statistical significance, resulting in 46,080 negotiations.

In this thesis, we do not study negotiation settings with reservation values other than zero (except for Chap. 9 on optimal bidding strategies with unknown reservation values). We do, however, employ agents and scenarios from ANAC 2012 throughout this thesis. When we do, we remove the reservation values from the preference specifications to ensure compatibility with the ANAC 2010 and 2011 agents.

B.3.4 ANAC 2013

ANAC 2013 had 19 participating teams from 8 different institutions. The local organizing committee responsible for ANAC 2013 was Ben Gurion University of the Negev, and the tournament was held during the Twelfth International Conference on Autonomous Agents and Multiagent Systems (AAMAS 2013) in Saint Paul, Minnesota, USA. The qualification round was played on 11 domains that were randomly selected from the submissions (see Table F.1). Each negotiation was repeated 10 times to establish statistical significance and to allow learning. Thus, every pair of agents played 20 times in each domain, totaling 75,240 negotiations.

The finals contained 7 agents, who were pitted on 18 different negotiation scenarios (12 submitted this year, plus 6 from 2012), which led to a total of 15120 negotiations. The ANAC 2013 domains contained between 1 and 7 issues, creating an outcome space of 3–56,700 possible outcomes (see Table F.3 and Figs. F.1 and F.2).

In 2013, we allowed agents to save information during and after negotiation session, and load it at the beginning of new session on the same domain and profile. Agents could use this information to learn about and adapt to the negotiation domains over time. As with the reservation values that were introduced in ANAC 2012, we do not explore this further in this thesis.

B.4 Conclusion

We described the goals and results of four installments of the International Auto-mated Negotiating Agents Competition (ANAC). The main purpose of ANAC is to motivate research in the area of bilateral multi-issue negotiations, with an emphasis on the practical design and development of successful automated negotiating agents. Additional goals include: collecting and objectively evaluating different state-of-the-art negotiation strategies and opponent models, defining a wide variety of bench-mark negotiation scenarios, and making them available to the negotiation research community.

Based on the submissions and the process of running the competition, as well as the post-tournament analysis reported in this thesis, we believe that this competition serves its purposes. The past competitions were successful events, enriching the research field on practical automated negotiation in line with the aims as set out in Sect. B.1. In particular, the widespread availability of efficient, general and domain-independent automated negotiators, which this tournament has achieved, has the advantages of minimizing the effort required for adaptation of a general automated negotiator to a new domain. Furthermore, the availability of the different agents allows researchers to have an objective measure to assist them in validating and testing the effectiveness of future automated negotiators.

One of the successes of ANAC lies in the development of state-of-the-art nego-tiation strategies that co-evolve every year. The four incarnations of ANAC already yielded more than 60 new strategies and scenarios which can be used as bench-marks to test the efficacy of subsequent work in this area, and we expect the trend of increasing participation to continue in the next years.

Not only can we learn from the strategy concepts introduced in ANAC, we also gain understanding in the correct setup of a negotiation competition, which in turn gives great insights into the deciding factors in the success of a negotiation agent.

The development of GENIUS is crucial to the organization of ANAC and con-versely, ANAC also advances the development of GENIUS. Moreover, GENIUS has proved itself as a valuable and extendable research and analysis tool for (post) tour-nament analysis. The success of ANAC underlines the importance of a flexible and versatile negotiation simulation environment such as GENIUS. GENIUS has the ability to run a wide range of different tournaments, an extensive repository of different agents and domains, and it contains standardized protocols and a scoring system, thus making it the perfect tournament platform for ANAC. Every year since 2010, we release a new, public build of GENIUS[10] containing all relevant aspects of ANAC. In particular, this includes all domains, preference profiles and agents that were used in the competition, in addition to the proposed improvements that were decided upon during the yearly discussions. Consequently, this makes it possible for the negotiation research community to do a complete re-run of ANAC and to perform subsequent in-depth analysis of other facets of negotiation encounters.

---

[10]http://ii.tudelft.nl/genius.

These appendices are based on the following publications: [15–17]

Tim Baarslag, Koen V. Hindriks, Catholijn M. Jonker, Sarit Kraus, and Raz Lin. The first automated negotiating agents competition (ANAC 2010). In Takayuki Ito, Minjie Zhang, Valentin Robu, Shaheen Fatima, and Tokuro Matsuo, editors, *New Trends in Agent-based Complex Automated Negotiations*, volume 383 of *Studies in Computational Intelligence*, pages 113–135, Berlin, Heidelberg, 2012. Springer-Verlag

Katsuhide Fujita, Takayuki Ito, Tim Baarslag, Koen V. Hindriks, Catholijn M. Jonker, Sarit Kraus, and Raz Lin. The second automated negotiating agents competition (ANAC 2011). In Takayuki Ito, Minjie Zhang, Valentin Robu, and Tokuro Matsuo, editors, *Complex Automated Negotiations: Theories, Models, and Software Competitions*, volume 435 of *Studies in Computational Intelligence*, pages 183–197. Springer Berlin Heidelberg, 2013

Tim Baarslag, Katsuhide Fujita, Enrico H. Gerding, Koen V. Hindriks, Takayuki Ito, Nicholas R. Jennings, Catholijn M. Jonker, Sarit Kraus, Raz Lin, Valentin Robu, and Colin R. Williams. Evaluating practical negotiating agents: Results and analysis of the 2011 international competition. *Artificial Intelligence*, 198:73–103, May 2013

Appendix C
ANAC 2010

In the Appendices C through F, we outline the main results of the four Automated Negotiating Agents Competitions (ANAC) that we organized between 2010 and 2013. We mainly focus on the participants, scenarios and results of the finals of each competition, with an emphasis on ANAC 2010 and 2011, since these are the results and resources most often used in this thesis.

The first installment of ANAC was in 2010 and was comprised of seven participating teams from five different universities, as listed in Table C.1.

The normalized domain scores of every agent in ANAC 2010 are listed in Table C.2. The normalized domain score is obtained by averaging the score against the other agents on multiple trials. All agents use both of the profiles that are linked to a domain. The final score is listed in the last column, thus making *Agent K* the winner of ANAC 2010.

Table C.1 Participating teams of ANAC 2010

Agent	Affiliation
IAMhaggler	University of Southampton
IAMcrazyHaggler	University of Southampton
Agent K	Nagoya Institute of Technology
Nozomi	Nagoya Institute of Technology
FSEGA	Babes Bolyai University
Agent Smith	Delft University of Technology
Yushu	University of Massachusetts Amherst

© Springer International Publishing Switzerland 2016
T. Baarslag, *Exploring the Strategy Space of Negotiating Agents*,
Springer Theses, DOI 10.1007/978-3-319-28243-5

Table C.2 Final scores and domain scores of every ANAC 2010 agent

Rank	Agent	Score per domain			Avg.
		Itex-Cyp	Eng–Zimb	Travel	
1	Agent K	0.901	0.712	0.685	0.766
2	Yushu	0.662	1.000	0.250	0.637
3	Nozomi	0.929	0.351	0.516	0.599
4	IAMhaggler	0.668	0.551	0.500	0.573
5	FSEGA	0.722	0.406	0	0.376
6	IAMcrazyHaggler	0.097	0.397	0.431	0.308
7	Agent Smith	0.069	0.053	0	0.041

C.1 Agents

We continue to report on the individual strategies of the ANAC 2010 agents, starting
with the winner. We compare the strategies by highlighting both common and contrasting approaches taken in the general strategic design. We are concerned with the
following aspects of proposal strategies:

1. **Proposal behavior**.
   For every agent, we give a brief overview of the basic decisions that comprise the
   agents' inner proposal loop. We also describe the criteria for accepting an offer.
   Either of the two can be decided in a deterministic or non-deterministic manner.
2. **Learning**.
   In order to reach an advantageous negotiation agreement, it is beneficial to have
   as much information about the preference profile of an opponent as possible. If
   an agent can take into consideration the opponent's interests and learn during
   their interactions, then their utility might increase [28]. Because of the closed
   negotiation setting of ANAC, the negotiating parties exchange only proposals,
   but they do not share any information about their preferences. To overcome this
   problem, a negotiating agent may try to obtain a model of the preference profile of
   its opponent by means of learning. For the participating agents, we are concerned
   how their strategies model the opponent.
3. **Timing aspects**.
   There are substantial risks associated with delaying the submission of a proposal
   at the end of the negotiation. These risks arise from unpredictable delays and can
   cause proposals to be received when the game is already over. Agents can try to
   estimate the length of their negotiation cycles to cope with these risks. The agents
   can then concede in the final phase of the negotiation, or place their proposals in
   some calculated amount of time before the end. We examine whether the agents
   make any predictions on how many time is left and how they use this information.

Agent K

The proposal mechanism of *Agent K* [29] works as follows: based on the previous proposals of the opponent and the time that is left, it sets a so-called *proposal target* (initially set to 1). If it already received an offer that matches at least the utility of the proposal target, it will offer this proposal to improve the chances of acceptance. Otherwise, it searches for random proposals that are at at least as good as the proposal target. If no such proposals are found, the proposal target is slightly lowered.

The agent has a sophisticated mechanism to accept an offer. It uses the mean and variance of the utility of all received offers, and then tries to determine the best offer it might receive in the future and sets its proposal target accordingly. It then accepts or rejects the offer, based on the probability that a better offer might be proposed. For more information and technical details on *Agent K*, see [29].

Yushu

Yushu [30] is a fairly simple agent that makes use of a target utility to make its next offer. As a learning mechanism, it uses the ten best proposals made by the opponent, called *suggested proposals*. It also makes an estimate of how many rounds are still left for the negotiation. Combining this information, *Yushu* obtains the target utility. It also keeps track of the acceptability-rate: the minimum utility it is willing to accept. To set the acceptability-rate, *Yushu* first finds the best possible utility that can be obtained in the domain, and accepts no less than 96 % of it. When the number of estimated future rounds becomes short, this percentage is lowered to 92 %.

The agent can only accept a proposal when the offered utility is above the target utility or when the utility reaches the acceptability-rate. Provided that either of the two is the case it accepts, when there are less than eight rounds left. When there is more time, it will accept only if it cannot find a suggested proposal with a better utility. If a better suggested proposal is available, it will offer that instead.

Nozomi

The proposal strategy of *Nozomi* [29] starts with an offer of maximum utility. It defines the gap between two parties as the differences in utility of their last offers. Depending on the gap and time that is left, it then chooses to make a certain proposal type, such as making a compromise, or staying put. *Nozomi* keeps track of the compromises made, but the agent does not model the utility function of the opponent.

The agent splits the negotiation into four intervals around 50, 80 and 90 % of the negotiation time. Based on previous offers, the gap between the two parties, and the time that is left in the negotiation, it will choose whether to accept an offer or reject it.

IAMHaggler and IAMcrazyHaggler

IAMhaggler and *IAMcrazyHaggler* (cf. [31]) are both implementations of a framework called *SouthamptonAgent*, thus creating a lot of similarity between the two agents. The *SouthamptonAgent* provides standard methods for handling offers, proposing offers and keeping track of time. The framework is the only one that also keeps track of the time that the opponent uses.

IAMcrazyHaggler is a very simple take-it-or-leave-it strategy: it will make random proposals with a utility that is above a constant threshold, set to 0.9 (without discount factors it is set to 0.95). The proposal is done without regard to time or opponent moves.

IAMHaggler, on the other hand, is a fully fledged negotiation strategy, which incorporates a model of the opponent using Bayesian learning. It starts with a proposal of maximum utility and successively sets a target utility based on multiple factors, such as: the utility offered by the opponent, the time left for both agents, and the perceived opponent's profile, such as hardheadedness. Upon receiving an offer, it analyzes the previous proposals of the opponent and adapts the hypotheses on the opponent's utility function. With this opponent model, it tries to find trade-offs that satisfy the target utility.

Let u be the utility of the last opponent's offer. Both agents accept an offer depending on u, namely when either of the following three conditions is met:

1. When u is at least 98 % of the utility of its own previous offer.
2. When u is at least 98 % of a *maximum aspiration* constant. The default value is 0.9, but if there are discount factors it is set to 0.85 for *IAMcrazyHaggler* to make it reach an agreement sooner.
3. When u is at least 98 % of the utility of its own upcoming offer.

Note that the three conditions only depend on the utility of the offer and not on the available time.

FSEGA

Similar to *Nozomi*, the *FSEGA* strategy [32] splits the negotiation into three intervals of time and applies different sub-strategies to each interval:

1. The first interval consists of the starting 85 % of the negotiation time and is mainly used to acquire the opponent's profile from the counter-offers.
2. In the next 10 %, the proposal strategy still does not concede, but relaxes some conditions for selecting the next proposal to improve the chances that the opponent accepts. The agent makes only small concessions and still tries to learn the opponent's profile.
3. In the final 5 %, *FSEGA* considers the time restrictions and employs a concession-based strategy to select the next offer up to its reservation value.

In the first phase of the negotiation, the accept mechanism will admit any opponent offer that is 3 % better than the utility of *FSEGA*'s last proposal. It will also always accept the best possible proposal. Otherwise, it selects a new proposal, but if the previous opponent's offer is better than the upcoming proposal it will accept it instead. After interval 1, it will also accept when it cannot find a better proposal for the opponent.

Agent Smith

Agent Smith [33] constructs an opponent model that represents the importance and preference for all values of each issue. The agent starts by making a first proposal of maximum utility and subsequently concedes slowly towards the opponent.

The agent accepts an offer given the following circumstances. The agents' threshold for acceptance slowly decreases over time. In the last 10 s of the negotiation session, *Agent Smith* will propose the best proposal that the opponent already proposed (even when the offer is very bad for itself). Since it previously proposed it, it is likely for a rational opponent to accept this proposal. However, an error was made in the implementation, resulting in the fact that the agent already shows this behavior after two minutes instead of three. This explains the poor performance of the agent in the competition.

C.2   Scenarios

Three scenarios were selected for ANAC 2010 by the organizers, which can be viewed in Table C.3.

England–Zimbabwe

The first scenario of ANAC 2010 is taken from [2, 26], which involves a case where *England* and *Zimbabwe* are negotiating to reach an agreement in response to the world's first public health treaty: the World Health Organization's Framework Convention on Tobacco Control. The leaders of both countries must reach an agreement on five issues:

Funding Amount The total amount to be deposited into a fund to aid countries that are economically dependent on tobacco production. This issue has a negative impact on the budget of *England* and a positive effect on the economy of *Zimbabwe*. The possible values are no agreement, $10, $50 or $100 billion. Thus, this issue has a total of four possible values.

Table C.3 Details of all ANAC 2010 scenarios

Domain	Size	Issues	Opposition	Bid distribution
England–Zimbabwe	576	5	0.278	0.298
Itex versus Cypress	180	4	0.431	0.222
Travel	188,160	7	0.230	0.416

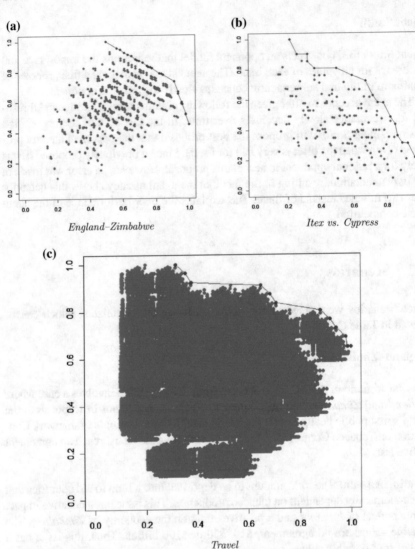

Fig. C.1 Outcome spaces of all ANAC 2010 scenarios. The points represent all of the outcomes that are possible in each scenario. The *solid line* is the Pareto frontier, which connects all of the Pareto efficient outcomes

Other Aid Programs The impact on other aid programs. If other aid programs are reduced, then this will create economic difficulties for *Zimbabwe*. Possible values are:

1. No reduction;
2. Reduction equal to half of the fund;

3. Reduction equal to the whole size of the fund;
4. No agreement.

Thus, a total of four possible values are allowed for this issue.

Trade Barriers Trade issues for both countries. *Zimbabwe* and *England* can use trade barriers such as tariffs (taxes on imports) or they can abstain from restrictive trade barriers to increase imports from the other party.

There is a trade-off in revenue of these policies: tariffs increases short-time revenue, but can lead to higher consumers prices. Decreasing import is good for local industries but it can decrease costumer welfare due to the increase in costumer costs. There are actually two issues here: the trade barriers that either side decides to use. *Zimbabwe*'s possible values are divided between

1. Reducing tariffs on imports;
2. Increasing tariffs on imports;
3. No agreement.

While *England* can choose between:

1. Reducing imports;
2. Increasing imports;
3. No agreement.

Thus, a total of three possible values are allowed for each of the two issues.

Creation of a Forum A forum can be created to explore other arrangements for health-issues. *Zimbabwe* would like to establish such a fund, to be able to apply to other global health agreements in the future, while this would be costly for *England*. The four possible values are:

1. Creation of a fund;
2. Creation of a committee that will discuss the creation of a fund;
3. Creation of a committee that will develop an agenda for future discussions;
4. No agreement.

Consequently, the domain has a total of $4^3 \times 3^2 = 576$ possible agreements. *England* and *Zimbabwe* have contradictory preferences for the first two issues, but the other issues have options that are jointly preferred by both sides, making it a domain of medium opposition.

Itex versus Cypress

The second scenario of ANAC 2010 is taken from [34], which describes a buyer–seller business negotiation for one commodity. It involves representatives of two companies: *Itex Manufacturing*, a producer of bicycle components and *Cypress Cycles*, a builder of bicycles. There are four issues that both sides have to discuss: *the price of* the components, delivery times, payment arrangements, and terms for the return of possibly defective parts. An example outcome would be:

$$(\$3.98, 45\ days,\ payment\ upon\ delivery, 5\%\ spoilage\ allowed).$$

The opposition is strong in this domain, as the manufacturer and consumer have naturally opposing needs and requirements. Altogether, there are 180 potential offers that contain all combinations of values for the four issues.

Travel

The final domain of ANAC 2010 has two persons negotiating to go on holiday to a location. From a small travel recommendation system we obtained multiple real-life profiles of travelers. They can each list their preferences on seven properties of a holiday destination: *Atmosphere, Amusement, Culinary, Shopping, Culture, Sport,* and *Environment.*

These properties determine the seven issues to discuss, all with a fairly large amount of choices. This leads to a big offers space of 188,160 possibilities. A sample negotiation outcome reads:

(*Hospitable, Nightlife and entertainment, International cuisine, Small boutiques, Art galleries, Outdoor activities, Parks and gardens*).

The opposition is weak in this domain, because traveling friends may have very compatible interests. Still the challenge is to find this optimal outcome in such a big search space.

Appendix D
ANAC 2011

Eighteen teams from seven different institutes and six different countries submitted negotiating agents to the 2011 tournament. The qualifying round results are the average over all 18 scenarios, which were submitted by the participants. Eight of these teams continued to the finals after undergoing a qualifying round (see Table D.1).

Table D.2 shows the scores of the 8 finalists of the tournament on the 8 scenarios submitted by all finalists.

D.1 Agents

In this section, we provide, in alphabetical order, brief descriptions of the individual strategies of the finalists of ANAC 2011 based on descriptions of the strategies provided by the teams.

Agent K2

This agent is identical to *Agent K* [29], winner of the ANAC 2010 competition. When creating a counter offer *Agent K* calculates a target utility U_t based on the previous offers made by the opponent and the time that is still remaining in the negotiation. *Agent K* then makes random bids above the target utility. If no such bid can be found, the target utility is lowered to allow for more offers. The target utility U_t at time t is calculated using the following formula:

$$U_t = 1 - (1 - E_{\max}(t)) \cdot t^\alpha, \tag{D.1}$$

where $E_{\max}(t)$ is the estimated maximum value the opponent will present in the future based on the average and variance of previous bids, and α is a parameter which controls the concession speed.

Agent K uses quite a sophisticated acceptance mechanism, where it will use the average and variations of the previous bid utilities presented by the opponent to deter-

© Springer International Publishing Switzerland 2016
T. Baarslag, *Exploring the Strategy Space of Negotiating Agents*,
Springer Theses, DOI 10.1007/978-3-319-28243-5

Table D.1 Scores and affiliation of every strategy in the qualifying round of ANAC 2011

Rank	Score	Agent	Affiliation
1	0.756	Gahboninho	Bar-Ilan University
2	0.708	HardHeaded	Delft University of Technology
3	0.706	ValueModelAgent	Bar-Ilan University
4	0.702	Agent K2	Nagoya Institute of Technology
5	0.701	IAMhaggler2011	University of Southampton
6	0.690	BRAMAgent	Ben-Gurion University
7	0.686	Nice Tit for Tat Agent	Delft University of Technology
8	0.685	The Negotiator	Delft University of Technology
9	0.678	GYRL	Ben-Gurion University
10	0.671	WinnerAgent	Ben-Gurion University
11	0.664	Chameleon	University Politehnica of Bucharest
12	0.648	SimpleAgentNew	Ben-Gurion University
13	0.640	LYYAgent	Bar-Ilan University
14	0.631	MrFriendly	Delft University of Technology
15	0.625	AgentSmith	Bar-Ilan University
16	0.623	IAMcrazyHaggler	University of Southampton
17	0.601	DNAgent	Universidad de Alcala
18	0.571	ShAgent	Bar-Ilan University

Table D.2 Tournament results in the final round of ANAC 2011

Rank	Agent strategy	Mean	Standard deviation	95 % Confidence interval	
				Lower bound	Upper bound
1	HardHeaded	0.749	0.0096	0.745	0.752
2	Gahboninho	0.740	0.0052	0.738	0.742
3	IAMhaggler2011	0.686	0.0047	0.685	0.688
4*	Agent K2	0.681	0.0047	0.679	0.683
5*	The Negotiator	0.680	0.0043	0.679	0.682
6*	BRAMAgent	0.680	0.0050	0.678	0.682
7*	Nice Tit for Tat Agent	0.678	0.0076	0.675	0.681
8	ValueModelAgent	0.617	0.0069	0.614	0.619

mine the best possible bid it can expect in the future. It will either accept or reject the offer based on the probability that the opponent will present a better offer in the future. If it has already received an offer from the opponent with the same utility or higher, it will offer that bid instead.

BRAMAgent

This agent uses opponent modeling in an attempt to propose offers which are likely to be accepted by the opponent. Specifically, its model of the opponent stores the frequency with which each value of each issue is proposed. This information is maintained only over the 10 most recent offers received from the opponent. Therefore, the first 10 offers *BRAMAgent* makes will be its preferred bid (the one which maximizes its utility), while it gathers initial data for its opponent model.

It also uses a time-dependent concession approach, which sets a threshold at a given time. In each turn, *BRAMAgent* tries to create a bid that contains as many of the opponent's preferred values as possible (according to its opponent model), with a utility greater than or equal to the current threshold. If *BRAMAgent* fails to create such a bid, a bid will be selected from a list of bids that was created at the beginning of the session. This list contains all of the possible bids in the scenario (or all the bids it managed to create in 2 s), sorted in descending order according to the utility values. *BRAMAgent* chooses randomly a bid that is nearby the previous bid that was made from that list.

BRAMAgent will accept any offer with utility greater than its threshold. The threshold, which affects both acceptance and proposal levels, varies according to time. Specifically, the threshold levels are set as pre-defined, fixed percentages of the maximum utility that can be achieved (0–60 s: 93% of the maximum utility, 60–150 s: 85%, 150–175 s: 70%, 175–180 s: 20%).

Gahboninho

This agent uses a meta-learning strategy that first tries to determine whether the opponent is trying to learn from its own concessions, and then exploits this behavior. Thus, during the first few bids, *Gahboninho* steadily concedes to a utility of 0.9 in an attempt to determine whether or not the opponent is trying to profile the agent. At the same time, the agent tries to assert selfishness and evaluate whether or not the opponent is cooperative. The degree of the opponent's selfishness is estimated based on the opponent's proposals. Then, the more the opponent concedes, the more competitive *Gahboninho*'s strategy becomes. The opponent's willingness to concede is estimated based on the size of variance of the opponent's proposals. After this phase, if the opponent is deemed concessive or adaptive, the agent takes a selfish approach, giving up almost no utility. However, if the opponent asserts even more hard-headedness, it adapts itself to minimize losses, otherwise it risks breakdown in the negotiation (which has very low utility for both parties). In generating the bids, the agent calculates its target, U_t at time t as follows:

$$U_t = U_{\max} - (U_{\max} - U_{\min}) \cdot t \tag{D.2}$$

where $U_{\max}$ and $U_{\min}$ are the maximum and minimum utilities (respectively) in the opponent's bidding history. $U_{\max}$ depends on the opponent's selfishness and the discount factor. Unlike many of the other agents, rather than using a model of the opponent to determine the offer to propose at a given utility level, *Gahboninho* uses a random search approach. Specifically, the agent proposes a random offer above the target utility $T(t)$. The benefit of this approach is that it is fast, therefore, given the format of the competition, a very large number of offers can be exchanged, allowing greater search of the outcome space. Moreover, the agent suggests using the opponent's best bid if the time is almost up.

HardHeaded

In each negotiation round, *HardHeaded* considers a set of bids within a pre-defined utility range which is adjusted over time by a pre-specified, monotonically decreasing function. A model of the opponent's utility function is constructed by analyzing the frequency of the values of the issues in every bid received from the opponent. From a set of bids with approximately equal utility for the agent itself, the opponent model is used to suggest bids that are best to the opponent in order to increase chances of reaching an agreement in a shorter period of time.

The concession function specifies an increasing rate of concession (i.e. decreasing utility) for the utility of the agent's bids. The function has non-monotonic curvature with one inflection point, determined by the discount factor of the scenario. This function is determined by tuning the strategy based on the sample scenarios and data made available before the competition. For the scenarios with time discounting, the timeline is split into two phases over which the agent practices different strategies: it starts by using a *Boulware* strategy, and after a certain amount of time has passed (depending on the discount factor), it switches to a *Conceder* strategy [35].

IAMhaggler2011

This agent uses a Gaussian process regression technique to predict the opponent's behavior [36]. It then uses this estimate, along with the uncertainty values provided by the Gaussian process, in order to optimally choose its concession strategy. In so doing, the concession strategy considers both the opponent's behavior and the time constraints.

The concession strategy is then used to determine the target utility at a given time. In the concession strategy, the agent finds the time, t^*, at which the expected discounted utility of the opponent's offer is maximized. In addition, it finds the utility level, u^*, at which the expected discounted utility of our offer is maximized. The agent then concedes towards $[t^*, u^*]$, whilst regularly repeating the Gaussian process and maximizations.

Finally, having chosen a target, the agent proposes an offer which has a utility close to that target. In choosing the bids, *IAMhaggler2011* uses an approach similar to that of *Gahboninho*. Specifically, a random package, with utility close to the target is selected according to the concession strategy. This strategy is a fast process, which allows many offers to be made and encourages the exploration of outcome space.

Nice Tit for Tat Agent

This agent plays a tit-for-tat strategy with respect to its own utility. The agent will initially cooperate, then respond in kind to the opponent's previous action, while aiming for the Nash point in the scenario. If the opponent's bid improves its utility, then the agent concedes accordingly. The agent is nice in the sense that it does not retaliate. Therefore, when the opponent makes an offer which reduces the agent's utility, the *Nice Tit for Tat Agent* assumes the opponent made a mistake and does nothing, waiting for a better bid. This approach is based on [37]. *Nice Tit for Tat Agent* maintains a Bayesian model [38] of its opponent, updated after each move by the opponent. This model is used to try to identify Pareto optimal bids in order to be able to respond to a concession by the opponent with a nice move. The agent will try to mirror the opponent's concession in accordance with its own utility function.

The agent detects very cooperative scenarios to aim for slightly more than Nash utility. Also, if the domain is large, if the discount factor is high, or if time is running out, the agent will make larger concessions towards its bid target. The agent tries to optimize the opponent's utility by making a number of different bids with approximately this bid target utility.

The Negotiator

Unlike the other finalist agents, this agent does not model the opponent. Its behavior depends on the mode it is using, which can be either: DISCOUNT or NODISCOUNT. A negotiation starts with the agent using its NODISCOUNT mode, which results in hardheaded behavior. After a predetermined time period, the agent switches to its DISCOUNT mode, in which its behavior becomes more concessive.

The main difference between the different modes is in the speed of descent of the minimum threshold for acceptance and offering. In the NODISCOUNT mode, most time is spent on the higher range of utilities and only in the last seconds are the remaining bids visited. The DISCOUNT mode treats all bids equally and tries to visit them all. An opponent's offer is accepted if it is above the current minimum threshold. An offer should also satisfy the minimum threshold, however a dynamic upper-bound is used to limit the available bids to offer in a turn. In 30% of the cases this upper-bound is ignored to revisit old bids, which can result in acceptance in later phases of the negotiation.

Finally, *The Negotiator* attempts to estimate the number of remaining moves to ensure that it always accepts before the negotiation deadline.

ValueModelAgent

This agent uses temporal difference reinforcement learning to predict the opponent's utility function. The particular learning technique is focused on finding the amount of utility lost by the opponent for each value. However, as the bid (expected) utilities represent the decrease in all issues, a method is needed to decide which values should change the most. To achieve this, the agent uses estimations of standard deviation and reliability of a value to decide how to make the split. The reliability is also used to decide the learning factor of the individual learning. The agent uses a symmetric lower-bound to approximate the opponent's concession (if the opponent makes 100 different bids, and the 100th bid is worth 0.94, it is assumed the opponent conceded at least 6 %). These parameters were determined in advance, based on average performance across a set of scenarios available for testing before the competition.

In more detail, *ValueModelAgent* starts by making bids which lie in the top 2 % of the outcome space. It severely limits the concession in the first 80 % of the timeline. If there is a large discount, the agent compromises only as much as its prediction of the opponent's compromise. If there is no discount, the agent does not concede as long as the opponent is compromising. If the opponent stops moving, the agent compromises up to two thirds of the opponent's approximated compromise. As the deadline approaches (80–90 % of the time has elapsed), the agent compromises up to 50 % of the difference, providing that the opponent is still not compromising. Once 90 % of the time has elapsed, the agent sleeps and makes the "final offer", if the opponent returns offers the agent sends the best offer that has been received from the opponent (accepting his last offer only if its close enough). *ValueModelAgent* has a fixed lower limit on its acceptance threshold, of 0.7. Therefore it never accepts an offer with an undiscounted utility lower than this value.

D.2   Scenarios

The properties of the 8 scenarios submitted by the finalists of ANAC 2011 are listed in Table D.3, and the shape of the outcome space of each scenario is presented graphically in Figs. D.1 and D.2.

Table D.3 Details of the ANAC 2011 scenarios

Domain	Size	Issues	Discount factor	Opposition	Bid distribution
Amsterdam	3024	6	1.000	0.223	0.254
Camera	3600	6	0.891	0.252	0.448
Car	15,625	6	1.000	0.095	0.136
Energy	390,625	8	1.000	0.448	0.149
Grocery	1600	5	0.806	0.191	0.492
Company acquisition	384	5	0.688	0.125	0.121
Laptop	27	3	0.424	0.178	0.295
Nice Or Die	3	1	1.000	0.991	0.000

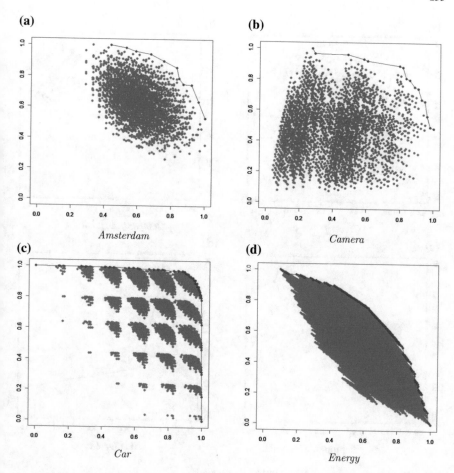

(a) *Amsterdam*

(b) *Camera*

(c) *Car*

(d) *Energy*

Fig. D.1 Outcome spaces of ANAC 2011 scenarios **a–d**

Nice Or Die

This scenario is the smallest used in the ANAC 2011 competition, with agents having to select between only 3 possible agreement points: a fair division point (nice), which is less efficient (in the sense that the sum of the agent's utilities is smaller) or one of two selfish points (die). The scenario is symmetric, in that neither player has an advantage over the other. The fair division point allows each player to achieve the same, relatively low score, while the other two selfish points allow one agent to get a high utility while its opponent achieves a very low one. As a result, the scenario has strong opposition between the participants. This means that if both agents try to get high utilities, it is hard for them to reach agreements. However, if agents would like to make an agreement in this scenario, the social welfare is small (as the agents cannot learn from previous interactions with an opponent).

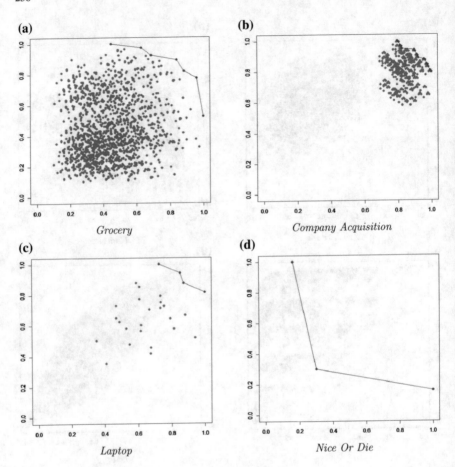

Fig. D.2 Outcome spaces of ANAC 2011 scenarios **e–h**

Laptop

In this scenario, a seller and a buyer are negotiating the specifications of a laptop. An agreement in the negotiation reconciles their differences and results in a purchase. The scenario has three issues: the laptop brand, the size of the hard disk, and the size of the external monitor. Each issue has only three options, making it a very small scenario with only 27 possible outcomes. Unbeknownst to each other, the buyer and seller actually both prefer to buy (and sell, respectively) a laptop with a small screen. The buyer prefers this because it is cheaper, and the seller prefers to sell laptops with small screens because s/he has more of those in stock. If the two parties are able to find the outcomes that are mutually beneficial to both, then they are happy to do business together with high utility scores on both sides.

Company Acquisition

This scenario represents a negotiation between two companies, in which the management of Intelligent Solutions Inc. (IS) wants to acquire the BI-Tech company (BT). The negotiation includes five issues: the price that IS pays for BI Tech, the transfer of intellectual property, the stocks given to the BI-Tech founders, the terms of the employees' contracts and the legal liability of Intelligent Solutions Inc. Each company wants to be the owner of the intellectual property. For IS, this issue is much more important. IS and BI-Tech have common interest that the BI-Tech co-founders would get jobs in IS. IS prefers to give BI-Tech only 2% of the stocks, while the BI-Tech co-founders want 5%. IS prefer private contracts, while firing workers is less desirable by them. BI-Tech prefers a 15% salary raise. For both sides this is not the most important issue in the negotiation. Each side prefers the least legal liability possible. In this case, the utility range is narrow and has high utility values such that all outcomes give both participants a utility of at least 0.5. The scenario is relatively small, with 384 possible outcomes.

Grocery

This scenario models a discussion in a local supermarket. The negotiation is between two people living together who have different tastes. The discussion is about five types of product: bread, fruit, snacks, spreads, and vegetables. Each category consists of four or five products, resulting in a medium sized scenario with 1,600 possible outcomes. For their daily routine it is essential that a product of each type is present in their final selection, however only one product can be selected for each type. Besides their difference in taste, they also differ in what category of product they find more important. The profiles for agents Mary and Sam are modeled in such a way that a good outcome is achievable for both. Sam has a slight advantage, since he is easier to satisfy than Mary, and therefore is likely to have better outcomes. This scenario allows outcomes that are mutually beneficial, but the outcome space is scattered so agents must explore it considerably to find the jointly profitable ones.

Amsterdam

This scenario concerns the planning of a tourist trip to Amsterdam and includes issues representing the day and time of travel, the duration of the trip, the type of venues to be visited, the means of transportation and the souvenirs to buy. This scenario is moderately large as the utility space has 3,024 possible bid configurations. The utility functions specify a generous win-win scenario, since it would be unrealistic for two friends to make a trip to Amsterdam and to have it be a zero-sum game. The size of the scenario enables the agent to communicate their preferences (by means of generating bids), without having to concede far. The size also puts agents which use a random method of generating bids at a disadvantage, since the odds of randomly selecting a Pareto optimal bid in a large scenario are small. So this scenario will give an advantage to agents that make some attempt to learn the opponents' utility function, and those capable of rapidly choosing offers.

Camera

This scenario is another retail based one, which represents the negotiation between a buyer and a seller of a camera. It has six issues: maker, body, lens, tripod, bags, and accessories. The size of this scenario is 3,600 outcomes. The seller gives priority to the maker, and the buyer gives priority to the lens. The opposition in this negotiation scenario is medium. The range of the contract space is wide, which means the agents need to explore it to find the jointly profitable outcomes. While jointly profitable outcomes are possible (since the Pareto frontier is concave) [12], no party has an undue advantage in this (since the Nash point is at an impartial position).

Car

This scenario represents a situation in which a car dealer negotiates with a potential buyer. There are 6 negotiation issues, which represent the features of the car (such as CD player, extra speakers and air conditioning) and each issue takes one of 5 values (good, fairly good, standard, meager, none), creating 15,625 possible agreements. Although the best bids of the scenario are worth zero for the opponent, this scenario is far from a zero-sum game. For example, agents can make agreements in which one of them can get close to the maximum possible utility, if it persuades its opponent to accept a utility only slightly below this. The scenario also allows agents to compromise to a fair division point in which both agents achieve a utility very close to the maximum possible. Consequently, the scenario has very weak opposition between the two participants.

Energy

This scenario considers the problem faced by many electricity companies to reduce electricity consumption during peak times, which requires costly resources to be available and puts a high pressure on local electricity grids. The application scenario is modeled as follows. One agent represents the electricity distribution company whilst the other represents a large consumer. The issues they are negotiating over represent how much the consumer is willing to reduce its consumption over a number of time slots for a day-ahead market (the 24 h in a day are discretized into 3 hourly time slots). For each issue, there is a demand reduction level possible from zero up to a maximum possible (specifically, 100 kW). In this scenario, the distributor obtains utility by encouraging consumers to reduce their consumptions. Participants set their energy consumption (in kWh) for each of 8 time slots. In each slot, they can reduce their consumption by 0, 25, 50, 75 or 100 kWh. This scenario is the largest in the 2011 competition (390,625 possible agreements) and has highly opposing utility functions, therefore, reaching mutually beneficial agreements requires extensive exploration of the outcome space by the negotiating agents.

Appendix E
ANAC 2012

For ANAC 2012, 17 teams entered from 8 different institutions from 5 countries. For the qualifying round, negotiations were carried out for every combination of the 17 participants on 18 negotiation scenarios (17 submitted this year, plus the *Travel* domain from 2010); see Table E.1.

Table E.1 Scores and affiliation of every strategy in the qualifying round of ANAC 2012

Rank	Score	Agent	Affiliation
1–2	0.597	*CUHK Agent*	The Chinese University of Hong Kong
1–2	0.590	*OMAC Agent*	Maastricht University
3–5	0.572	*The Negotiator Reloaded*	Delft University of Technology
3–7	0.568	*BRAMAgent2*	Ben-Gurion University
3–7	0.565	*Meta-Agent*	Ben-Gurion University
4–7	0.564	*IAMhaggler2012*	University of Southampton
4–8	0.563	*AgentMR*	Nagoya Institute of Technology
8–10	0.550	*AgentLG*	Bar-Ilan University
7–9	0.556	*IAMcrazyHaggler2012*	University of Southampton
9–11	0.547	*Agent Linear*	Nagoya Institute of Technology
10–11	0.542	*Rumba*	Bar-Ilan University
12	0.521	*Dread Pirate Roberts*	Delft University of Technology
13–14	0.469	*AgentX*	Delft University of Technology
13–14	0.465	*AgentI*	Nagoya Institute of Technology
15–16	0.455	*AgentNS*	Nagoya Institute of Technology
15–16	0.447	*AgentMZ*	Nagoya Institute of Technology
17	0.394	*AgentYTY*	Shizuoka University

© Springer International Publishing Switzerland 2016
T. Baarslag, *Exploring the Strategy Space of Negotiating Agents*,
Springer Theses, DOI 10.1007/978-3-319-28243-5

Table E.2 Final ranking of every strategy in ANAC 2012

Rank	Agent	Score	Variance
1	CUHK Agent	0.626	0.000003
2	AgentLG	0.622	0.000003
3–4	OMAC Agent	0.618	0.000002
3–4	The Negotiator Reloaded	0.617	0.000002
5	BRAMAgent2	0.593	0.000002
6	Meta-Agent	0.586	0.000003
7	IAMhaggler2012	0.535	0.000001
8	AgentMR	0.328	0.000003

For the final round, 8 agents were selected, together with 24 base scenarios (17 submitted this year, 5 from 2011, and 2 from 2010). The results of the final round are shown in Table E.2.

E.1 Scenarios

There were 13 scenarios that featured for the first time in ANAC 2012, which are shown in Table E.3 (Figs. E.1, E.2 and E.3).

Table E.3 Details of the ANAC 2012 scenarios

Domain	Size	Issues	Opposition	Bid distribution
Airport Site Selection	420	3	0.296	0.361
Barbecue	1440	5	0.248	0.277
Barter	80	3	0.492	0.036
Energy Small	15,625	6	0.432	0.217
Fifty-Fifty	11	1	0.707	0.000
Fitness	3520	5	0.275	0.283
Flight Booking	48	3	0.326	0.166
House Keeping	384	5	0.281	0.239
Music Collection	4320	6	0.158	0.343
Outfit	128	4	0.198	0.327
Phone	1600	5	0.194	0.490
Rental House	60	4	0.327	0.096
Supermarket	112,896	6	0.347	0.347

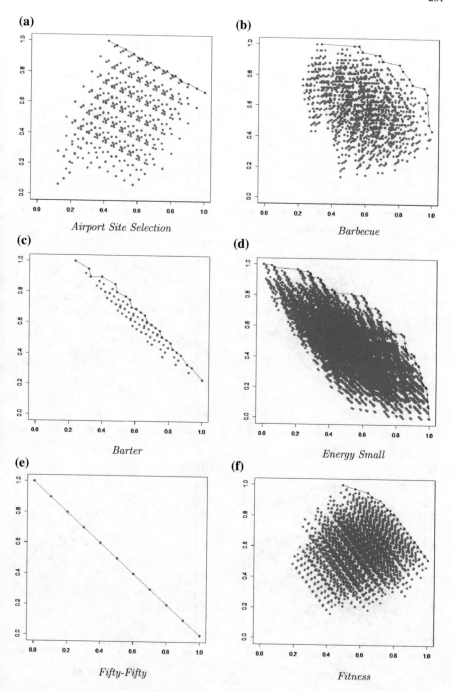

Fig. E.1 Outcome spaces of ANAC 2012 scenarios **a–f**

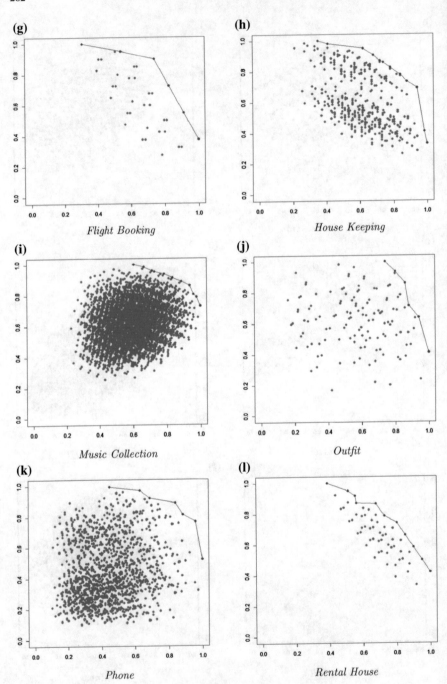

Fig. E.2 Outcome spaces of ANAC 2012 scenarios **g–l**

Fig. E.3 The *Supermarket* outcome space of ANAC 2012

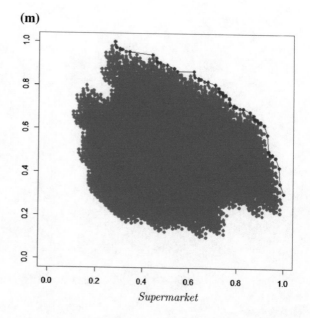

(m)

Supermarket

Appendix F
ANAC 2013

ANAC 2013 had 19 participating teams from 8 different institutions (see Table F.1). The qualification round was played on 11 domains that were randomly selected from the submissions.

Table F.1 Scores and affiliation of every strategy in the qualifying round of ANAC 2013

Rank	Score	Agent	Affiliation
1	0.562	Agent KF	Tokyo University of Agriculture and Technology
2–3	0.522	The Fawkes	Delft University of Technology
2–4	0.516	TMF Agent	Ben-Gurion University
3–4	0.495	Meta-Agent	Ben-Gurion University
5–8	0.457	G-Agent	Nagoya institute of technology
5–8	0.455	Inox Agent	Delft University of Technology
5–11	0.447	Slava Agent	Bar-Ilan University
5–11	0.446	VAStockMarketAgent	Ben-Gurion University
7–11	0.432	RoOAgent	Shizuoka University
7–11	0.431	Agent Talex	Ben-Gurion University
7–11	0.43	AgentMRK2	Nagoya Institute of Technology
12–14	0.387	Elizabeth	Nagoya Institute of Technology
12–15	0.374	ReuthLiron	Ben-Gurion University
12–15	0.373	BOA Constrictor	Delft University of Technology
13–18	0.359	Pelican	Nagoya Institute of Technology
15–18	0.35	Oriel Einat Agent	Ben-Gurion University
15–18	0.345	Master Qiao	Maastricht University
15–18	0.338	E Agent	Nagoya Institute of Technology
19	0.315	Clear Agent	Bar-Ilan University

© Springer International Publishing Switzerland 2016
T. Baarslag, *Exploring the Strategy Space of Negotiating Agents*,
Springer Theses, DOI 10.1007/978-3-319-28243-5

Table F.2 Final ranking of every strategy in ANAC 2013

Rank	Agent	Score	Variance
1	*The Fawkes*	0.606434	0.000011
2	*Meta-Agent*	0.600209	0.000083
3	*TMF Agent*	0.583094	0.000012
4–5	*Inox Agent*	0.568215	0.000069
4–5	*G-Agent*	0.564908	0.000055
6	*Agent KF*	0.534514	0.000147
7	*Slava agent*	0.484973	0.000023

The finals contained 7 agents, who were pitted on 18 different negotiation scenarios (12 submitted this year, plus 6 from 2012). The results of the final round are shown in Table F.2.

Note that we won first place in the ANAC 2013 competition with *The Fawkes*, an agent that used the BOA architecture to combine several components that were known to be effective at the time.[11] The bidding strategy and opponent modeling component are based on the *OMAC Agent* [39] from ANAC 2012. To improve the bidding strategy, the agent was designed to be more generous as the time passes instead of using a fixed target utility range. For the acceptance mechanism, we selected a version of $AC_{combi}(T, AVG^W)$, which is shown to be among the most effective acceptance mechanisms in Chap. 4.

F.1 Scenarios

The ANAC 2013 domains contained between 1 and 7 issues, creating an outcome space of 3 to 56,700 possible outcomes (see Table F.3).

---

[11]This means some of our later insights were not used in its design, such as our results on optimal stopping (Chap. 5) and optimal bidding (Chap. 9).

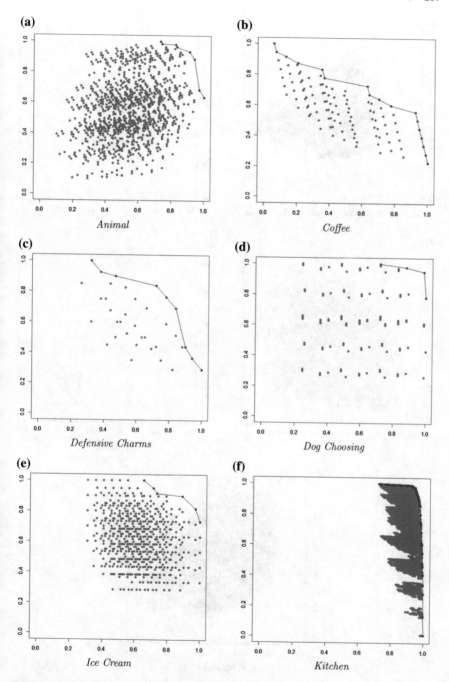

Fig. F.1 Outcome spaces of ANAC 2013 scenarios **a–f**

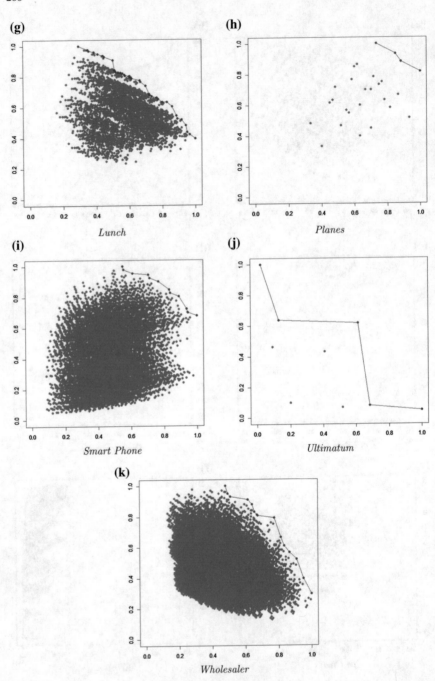

Fig. F.2 Outcome spaces of ANAC 2013 scenarios **g–k**

Table F.3 Details of the ANAC 2013 scenarios

Domain	Size	Issues	Opposition	Bid distribution
Animal	1152	5	0.110	0.429
Coffee	112	3	0.486	0.145
Defensive charms	36	3	0.322	0.165
Dog Choosing	270	5	0.051	0.471
Ice Cream	720	4	0.148	0.328
Kitchen	15,625	6	0.063	0.071
Lunch	3840	6	0.420	0.196
Planes	27	3	0.165	0.311
Smart Phone	12,000	6	0.237	0.512
Ultimatum	9	2	0.545	0.123
Wholesaler	56,700	7	0.308	0.394

Summary

Negotiation is an important activity in human society, and is studied by various disciplines, ranging from economics and game theory, to electronic commerce, social psychology, and artificial intelligence. Traditionally, negotiation is a necessary, but also time-consuming and expensive activity. Therefore, in the last decades there has been a large interest in the *automation* of negotiation, for example in the setting of e-commerce. This interest is fueled by the promise of automated agents eventually being able to negotiate on behalf of human negotiators.

Every year, automated negotiation agents are improving in various ways, and there is now a large body of negotiation strategies available, all with their unique strengths and weaknesses. For example, some agents are able to predict the opponent's preferences very well, while others focus more on having a sophisticated bidding strategy. The problem however, is that there is little incremental improvement in agent design, as the agents are tested in varying negotiation settings, using a diverse set of performance measures. This makes it very difficult to meaningfully compare the agents, let alone their underlying techniques. As a result, we lack a reliable way to pinpoint the most effective components in a negotiating agent.

There are two major advantages of distinguishing between the different components of a negotiating agent's strategy: first, it allows the study of the behavior and performance of the components in isolation. For example, it becomes possible to compare the preference learning component of all agents, and to identify the best among them. Second, we can proceed to *mix and match* different components to create new negotiation strategies., e.g.: replacing the preference learning technique of an agent and then examining whether this makes a difference. Such a procedure enables us to combine the individual components to systematically explore the space of possible negotiation strategies.

The BOA Architecture

To develop a compositional approach to evaluate and combine the components, we identify structure in most agent designs by introducing the *BOA architecture* (Chap. 3), in which we can develop and integrate the different components of a negotiating agent. We identify three main components of a general negotiation strategy; namely a *bidding strategy* (B), possibly an *opponent model* (O), and an *acceptance strategy* (A). The *bidding strategy* considers what concessions it deems appropriate given its own preferences, and takes the opponent into account by using an *opponent model*. The *acceptance strategy* decides whether offers proposed by the opponent should be accepted.

The BOA architecture is integrated into a generic negotiation environment called GENIUS (Appendix A), which is a software environment for designing and evaluating negotiation strategies. To explore the negotiation strategy space of the negotiation research community, we amend the GENIUS repository with various existing agents and scenarios from literature. Additionally, we organize a yearly international negotiation competition (ANAC) (Appendix B) to harvest even more strategies and scenarios. ANAC also acts as an evaluation tool for negotiation strategies, and encourages the design of negotiation strategies and scenarios.

We re-implement agents from literature and ANAC and *decouple* them to fit into the BOA architecture without introducing any changes in their behavior. For each of the three components, we manage to find and analyze the best ones for specific cases, as described below. We show that the BOA framework leads to significant improvements in agent design by wining ANAC 2013, which had 19 participating teams from 8 international institutions, with an agent that is designed using the BOA framework and is informed by a preliminary analysis of the different components.

Acceptance Strategies

In every negotiation, one of the negotiating parties must accept an offer to reach an agreement. Therefore, it is important that a negotiator employs a proficient mechanism to decide under which conditions to accept. When contemplating whether to accept an offer, the agent is faced with *the acceptance dilemma*: accepting the offer may be suboptimal, as better offers may still be presented before time runs out. On the other hand, accepting too late may prevent an agreement from being reached, resulting in a break off with no gain for either party. In Chap. 4, we *classify* and *compare* state-of-the-art generic acceptance conditions. We propose new acceptance strategies and we demonstrate that they outperform the other conditions. We also provide insight into why some conditions work better than others and investigate correlations between the properties of the negotiation scenario and the efficacy of acceptance conditions.

In Chap. 5, we adopt a more principled approach by applying optimal stopping theory to calculate the optimal decision on the acceptance of an offer. We approach the decision of whether to accept as a *sequential decision problem*, by modeling the bids received as a stochastic process. We determine the *optimal acceptance policies* for particular opponent classes and we present an approach to estimate the expected range of offers when the type of opponent is unknown. We show that the proposed approach is able to find the optimal time to accept, and improves upon all existing acceptance strategies.

Opponent Models

Another principal component of a negotiating agent's strategy is its ability to take the *opponent's preferences* into account. The quality of an opponent model can be measured in two different ways. One is to use the agent's *performance* as a benchmark for the model's quality. In Chap. 6, we *evaluate* and *compare* the performance of a selection of state-of-the-art opponent modeling techniques in negotiation. We provide an overview of the factors influencing the quality of a model and we analyze how the performance of opponent models depends on the negotiation setting. We identify a class of simple and surprisingly effective opponent modeling techniques that did not receive much previous attention in literature.

The other way to measure the quality of an opponent model is to directly evaluate its *accuracy* by using *similarity measures*. We consider opponent models from this perspective in Chap. 7. We review all methods to measure the accuracy of an opponent model and we then analyze how changes in accuracy translate into performance differences. Moreover, we pinpoint the best *predictors* for good performance. This leads to new insights concerning how to construct an opponent model, and what we need to measure when optimizing performance.

Bidding Strategies

Finally, we take two different approaches to gain more insight into effective *bidding strategies*. In Chap. 8, we present a new classification method for negotiation strategies, based on their pattern of concession making against different kinds of opponents. We apply this technique to classify some well-known negotiating strategies, and we formulate guidelines on how agents should bid in order to be successful, which gives insight into the bidding strategy space of negotiating agents.

We focus on finding *optimal* bidding strategies in Chap. 9. We apply optimal stopping theory again, this time to find the concessions that maximize utility for the bidder against particular opponents. We show there is an interesting connection between optimal bidding and optimal acceptance strategies, in the sense that they are mirrored versions of each other.

Putting the Pieces Together

Lastly, after analyzing all components separately, we put the pieces back together again in Chap. 10. We take all BOA components accumulated so far, including the best ones, and combine them all together to explore the space of negotiation strategies.

We compute the contribution of each component to the overall negotiation result, and we study the interaction between components. We find that combining the best agent components indeed makes the strongest agents. This shows that the component-based view of the BOA architecture not only provides a useful basis for developing negotiating agents but also provides a useful analytical tool. By varying the BOA components we are able to demonstrate the contribution of each component to the negotiation result, and thus analyze the significance of each. The bidding strategy is by far the most important to consider, followed by the acceptance conditions and finally followed by the opponent model.

Our results validate the analytical approach of the BOA framework to first optimize the individual components, and then to recombine them into a negotiating agent.

References

1. Baarslag T (2013) Designing an automated negotiator: learning what to bid and when to stop. Proceedings of the 2013 international conference on autonomous agents and multi-agent systems, AAMAS '13. International Foundation for Autonomous Agents and Multiagent Systems, Richland, pp 1419–1420
2. Raz L, Sarit K, Jonathan W, James B (2008) Negotiating with bounded rational agents in environments with incomplete information using an automated agent. Artif Intell 172(6–7):823–851
3. Ficici SG, Pfeffer A (2008) Modeling how humans reason about others with partial information. Proceedings of the 7th international joint conference on autonomous agents and multiagent systems, AAMAS '08, vol 1. International Foundation for Autonomous Agents and Multiagent Systems, Richland, pp 315–322
4. Gal Y, Grosz BJ, Kraus S, Pfeffer A, Shieber S (2005) Colored trails: a formalism for investigating decision-making in strategic environments. In: Proceedings of the 2005 IJCAI workshop on reasoning, representation, and learning in computer games, pp 25–30
5. Gal Y, Grosz B, Kraus S, Pfeffer A, Shieber S (2010) Agent decision-making in open mixed networks. Artif Intell 174(18):1460–1480
6. Fabregues A, Navarro D, Serrano A, Sierra C (2010) DipGame: a testbed for multiagent systems. Proceedings of the 9th international conference on autonomous agents and multiagent systems, AAMAS '10, vol 1. International Foundation for Autonomous Agents and Multiagent Systems, Richland, pp 1619–1620
7. Fabregues A, Sierra C (2011) DipGame: a challenging negotiation testbed. Eng Appl Artif Intell 24(7):1137–1146
8. Kraus S, Lehmann D (1995) Designing and building a negotiating automated agent. Comput Intell 11(1):132–171
9. Wellman MP, Amy G, Peter S (2007) Autonomous bidding agents: strategies and lessons from the trading agent competition. MIT Press, Cambridge
10. Hindriks KV, Jonker CM (2009) Creating human-machine synergy in negotiation support systems: towards the pocket negotiator. In: Proceedings of the 1st international working conference on human factors and computational models in negotiation, HuCom '08. ACM, New York, pp 47–54
11. Raz L, Yehoshua G, Sarit K (2011) Bridging the gap: face-to-face negotiations with an automated mediator. IEEE Intell Syst 26(6):40–47
12. Howard R (1982) The art and science of negotiation: How to resolve conflicts and get the best out of bargaining. Harvard University Press, Cambridge
13. Hindriks KV, Jonker CM, Tykhonov D (2007) Negotiation dynamics: analysis, concession tactics, and outcomes. Proceedings of the 2007 IEEE/WIC/ACM international conference on intelligent agent technology, IAT '07. IEEE Computer Society, Washington, pp 427–433
14. Hindriks KV, Tykhonov D (2010) Towards a quality assessment method for learning preference profiles in negotiation. In: Wolfgang K, Johannes ALP, Norman S, Onn S, William W (eds) Agent-mediated electronic commerce and trading agent design and analysis, vol 44., Lecture notes in business information processingSpringer, Berlin, pp 46–59
15. Baarslag T, Hindriks KV, Jonker CM, Kraus S, Lin R, (2012) The first automated negotiating agents competition (ANAC, (2010) In: Ito T, Zhang M, Robu V, Fatima S, Matsuo T (eds) New

trends in agent-based complex automated negotiations, studies in computational intelligence, vol 383. Springer, Berlin, pp 113–135

16. Fujita K, Ito T, Baarslag T, Hindriks KV, Jonker CM, Kraus S, Lin R, (2013) The second automated negotiating agents competition (ANAC, (2011) In: Ito T, Zhang M, Robu V, Matsuo T (eds) Complex automated negotiations: theories, models, and software competitions, studies in computational intelligence, vol 435. Springer, Berlin, pp 183–197

17. Baarslag T, Fujita K, Gerding EH, Hindriks KV, Ito T, Jennings NR, Jonker CM, Kraus S, Lin R, Robu V, Williams CR (2013) Evaluating practical negotiating agents: Results and analysis of the 2011 international competition. Artif Intell 198:73–103

18. Baarslag T, Aydoğan R, Hindriks KV, Fuijita K, Ito T, Jonker CM (2015) The automated negotiating agents competition (ANAC) 2010–2015. AI Magazine 36

19. Williams CR, Robu V, Gerding EH, Jennings NR (2014) An overview of the results and insights from the third automated negotiating agents competition (ANAC (2012) In: Marsa-Maestre I, Lopez-Carmona MA, Ito T, Zhang M, Bai Q, Fujita K (eds) Novel insights in agent-based complex automated negotiation, studies in computational intelligence, vol 535. Springer, Japan, pp 151–162

20. Wellman MP, Wurman PR, Kevin O, Roshan B, de Lin S, Reeves D, Walsh WE (2001) Designing the market game for a trading agent competition. IEEE Internet Comput 5(2):43–51

21. Peyman F, Carles S, Jennings NR (1998) Negotiation decision functions for autonomous agents. Robot Auton Syst 24(3–4):159–182

22. Peyman F, Carles S, Jennings NR (2002) Using similarity criteria to make issue trade-offs in automated negotiations. Artif Intell 142(2):205–237

23. Ito T, Hattori H, Klein M (2007) Multi-issue negotiation protocol for agents: exploring nonlinear utility spaces. Proceedings of the 20th international joint conference on artifical intelligence, IJCAI'07. Morgan Kaufmann Publishers Inc, San Francisco, pp 1347–1352

24. Jonker CM, Valentin R, Jan T (2007) An agent architecture for multi-attribute negotiation using incomplete preference information. Auton Agents Multi-Agent Syst 15:221–252

25. Lin R, Oshrat Y, Kraus S (2009) Investigating the benefits of automated negotiations in enhancing people's negotiation skills. AAMAS '09: Proceedings of The 8th international conference on autonomous agents and multiagent systems. International Foundation for Autonomous Agents and Multiagent Systems, Richland, pp 345–352

26. Lin R, Kraus S, Tykhonov D, Hindriks KV, Jonker CM (2011) Supporting the design of general automated negotiators. In: Proceedings of the second international workshop on agent-based complex automated negotiations (ACAN'09), vol 319. Springer, pp 69–87

27. Rogers AD, Dash RK, Ramchurn SD, Perukrishnen V, Jennings Nicholas R (2007) Coordinating team players within a noisy iterated prisoner's dilemma tournament. Theor Comput Sci 377(1–3):243–259

28. Zeng D, Sycara KP (1997) Benefits of learning in negotiation. In: Proceedings of the fourteenth national conference on artificial intelligence and ninth conference on innovative applications of artificial intelligence, AAAI'97/IAAI'97. AAAI Press, pp 36–41

29. Kawaguchi S, Fujita K, Ito T (2012) Compromising strategy based on estimated maximum utility for automated negotiating agents. In: Ito T, Zhang M, Robu V, Fatima S, Matsuo T (eds) New trends in agent-based complex automated negotiations, series of studies in computational intelligence. Springer, Berlin, pp 137–144

30. An B, Lesser VR (2012) Yushu: a heuristic-based agent for automated negotiating competition. In: Ito T, Zhang M, Robu V, Fatima S, Matsuo T (eds) New trends in agent-based complex automated negotiations, studies in computational intelligence, vol 383. Springer, Berlin, pp 145–149

31. Williams CR, Robu V, Gerding EH, Jennings NR (2012) Iamhaggler: a negotiation agent for complex environments. In: Ito T, Zhang M, Robu V, Fatima S, Matsuo T (eds) New trends in agent-based complex automated negotiations, studies in computational intelligence. Springer, Berlin, pp 151–158

32. Şerban LD, Silaghi GC, Litan CM (2012) AgentFSEGA–time constrained reasoning model for bilateral multi-issue negotiations. In: Ito T, Zhang M, Robu V, Fatima S, Matsuo T (eds)

New trends in agent-based complex automated negotiations, series of studies in computational intelligence. Springer, Berlin, pp 159–165

33. van Galen Last N (2012) Agent Smith: opponent model estimation in bilateral multi-issue negotiation. In: Ito T, Zhang M, Robu V, Fatima S, Matsuo T (eds) New trends in agent-based complex automated negotiations, studies in computational intelligence. Springer, Berlin, pp 167–174

34. Gregory KE, Zhang G (2003) Mining inspire data for the determinants of successful internet negotiations. Central Eur J Oper Res 11(3):297–316

35. Fatima SS, Wooldridge MJ, Jennings NR (2002) Optimal negotiation strategies for agents with incomplete information. Revised papers from the 8th international workshop on intelligent agents VIII, ATAL '01. Springer, London, pp 377–392

36. Williams CR, Robu V, Gerding EH, Jennings NR (2011) Using gaussian processes to optimise concession in complex negotiations against unknown opponents. In: Proceedings of the twenty-second international joint conference on artificial intelligence, IJCAI'11, vol 1. AAAI Press, pp 432–438

37. Hindriks KV, Jonker CM, Tykhonov D (2009) The benefits of opponent models in negotiation. In: Proceedings of the 2009 IEEE/WIC/ACM international joint conference on web intelligence and intelligent agent technology, vol 2. IEEE Computer Society, pp 439–444

38. Hindriks KV, Tykhonov D (2008) Opponent modelling in automated multi-issue negotiation using bayesian learning. Proceedings of the 7th international joint conference on Autonomous agents and multiagent systems, AAMAS '08, vol 1. International Foundation for Autonomous Agents and Multiagent Systems, Richland, pp 331–338

39. Chen S, Weiss G (2012) An efficient and adaptive approach to negotiation in complex environments. In: de Raedt L, Bessiere C, Dubois D, Doherty P, Frasconi P, Heintz F, Lucas PJF (eds) ECAI, Frontiers in artificial intelligence and applications, vol 242. IOS Press, pp 228–233

40. Lin R, Kraus S, Baarslag T, Tykhonov D, Hindriks KV, Jonker CM (2014) Genius: An integrated environment for supporting the design of generic automated negotiators. Comput Intell 30(1):48–70

Printed in the United States
By Bookmasters